BRAIN FUNCTIONING AND REGENERATION

BRAIN FUNCTIONING AND REGENERATION

KINEMATICS OF THE BRAIN ACTIVITIES

VOLUME IV

MOSTAFA M. DINI

Copyright © 2012 by Mostafa M. Dini.

Library of Congress Control Number:		2012906854
ISBN:	Hardcover	978-1-4691-2607-4
	Softcover	978-1-4691-2606-7
	Ebook	978-1-4691-2608-1

All rights reserved. No part of this book may be reproduced or transmitted in any form or by any means, electronic or mechanical, including photocopying, recording, or by any information storage and retrieval system, without permission in writing from the copyright owner.

This book was printed in the United States of America.

To order additional copies of this book, contact:
Xlibris Corporation
1-888-795-4274
www.Xlibris.com
Orders@Xlibris.com

CONTENTS

Preface 9
Introduction 11

PART I

What Makes the Brain Different? 15
More about Sensory Input and Stress 16
Strain-stress effects in course of a brain activity 17
What is a Pathway? 18
Circumstances and Emotions 19
Creations in the Mind 20
Applications of the Dream Analysis 23
Emotional health 25

PART II

Brain Activity Parameters and Processes 27
Needs and Attractors 29
Stress flow Direction 31
Defining the Free Energy Flow Attached
   to a Brain Activity 34
Stress Flow in Calculation Terms 36
Resonations Between Brain Regions 38
The Required Activation Energy of Substrates 40
Stress Flow and Simultaneous Release of Strains 43
Strain-Stress Relationship and
   Absorption-Desorption Process 47

Experimental Stress-Strain Studies and
   the Kinematic Model 49
Free Energy Transfer Phenomena and Similar
   Functional Patterns in Different Circumstances 52
Daily Strain Changes and Resulting Firing
   Gate Distribution Changes.. 55
Energy Wakes Forms and Discharges through
   the Outer Layer 61
Finding a sense of free energy flow over the layers 62
Measuring Energy Rates of Free Energy
   Transfers along the Layers 62

PART III

Input decomposition into static, dynamic and
   circumstantial evaluation 63
Building Up a Sense, Developing a Common Sense 65
Used, Experienced, and Practiced Patterns 66
Memory Consolidation 68
Mapping Memory Locations in the Brain............................ 70
Memory and its Physical Stress-Strain Basis...................... 71
Specific Family Memory Zone 72
Copy-Making Process of Energy Wakes in
   the Middle Brain 73
Structure of Memory Consolidation 74
Mapping Memory Locations Over the Brain...................... 75
Specific Family Memory Zone 76
Same Functionality Site for Different
   Character-Scenario Cases 77

PART IV

Interpreting of Detail Patterns with General
    Patterns in Dreams 77
Dream Automade Narrative 80
Dream Fictions 81
Symbolic Expression in the Dreams 83
Reality and Thoughts 84
Dreaming Principles 85
The Logic Behind Dreaming 86
A Dream Package Design 92
The Emotion Generator or Code Producer 94
Mechanism of Dreaming 97
Dream's Rules 99
A Dream as a Play 101
Characters' Images in Mathematic Description 104
Distinctions Between Characters and Play 105
Cognition of Static, Motional, Circumstantial, and
    Emotional Images 107
The Combined Pattern of the Releasing Networks
    Create the Overall Pattern of a Dream 109
Remembering a Dream 112

PART V

A Circumstance, Brain Processing, and the
    Emotion Involved 112
The Song of an Emotion 114
Emotional, Static, and Dynamic Elements
    in a Mentation 115

PART VI

Recommendation for the Basic Programming
 of the Dream Monitoring Package 116
Supporting project 119
The criteria for recording and analyzing the dream: 121

APPENDIXES:

Appendix A 123
Appendix B: An Emotion Tree (as an example) 129
Index 133

Preface

This book is the fourth volume of the book *Brain Functioning and Regeneration* and is written as a basis for a programming project for dream analysis, DreamWorks, and the production of a dream virtual monitoring software.

I would like to emphasize again that many statements in this text are claims and still not approved. However, as a model, they are essential to complete a detailed frame for programming intentions. The claims will be counterchecked with latest researches and will be refined for the noncontingencies continuously. I hope that in the near future, it helps to develop the unknown areas in the subject as well as provide an advanced software that is unique in its subject and services.

The next volumes of the book will be concentrating on comparisons between neuroscience and kinematics views to brain-function modeling.

I would like to thank my family for the support. I would like to thank Edmund Nolan, Victor Aushev, and Blaine White for assisting me in different aspects of the work.

Introduction

I had most of my work experience doing the conceptual design mostly in petrochemical and more specifically in polymer projects.

A few years back, I read some articles regarding the melt polymer behaviors when it would be pressed in one of the endings or any location along the containing pipe. The shear stresses imposed on the surface of the melt polymer plug and its time-dependence behavior became a subject of my interest.

Later, I read the *brain* is also a *viscoelastic material*, and I thought there should be some similarity between brain behaviors and other viscoelastic materials regarding straining and releasing memories, considering that the brain is exposed to *internal shear stresses* when the synapses and neuron fibers in a pathway are excited.

I also read brain neurons are grown in a fractal shape, and when the firings become synchronized in a pathway, the brain can be considered as a chaotic media developing fractals of firings through firing gates along the synapses in neural networks and the pathway made of those neural networks.

I thought any firing fractal of *synchronized firings* in a substrate of a *network* may be considered as an *energy packet,* and a *pathway* also can be considered as a bed for energy packets flowing from initiation by an input up to termination of that specific brain activity as an output.

Then I reached to the conclusion that any brain activity can be studied from a kinematic point of view in a way that *physics of elasticity* and *free energy transfer* through a bed of neural networks can be applied.

Based on these thoughts, I developed the subject of "Brain Functioning and Regeneration" and tried to search if there is any discrepancy between the predictions of this model and the findings of neuroscience. I found almost no discrepancies between neuroscience findings (which are being discussed in microscale) and my kinematic model predictions (which explain behaviors in macroscale).

Based on my studies, three previous volumes of the book were more concentrated on the principles of the kinematic model, and this one is searching to find applications in the possible ways. For example, defining a project to produce a software program that helps to map a spectrum of emotions during an individual life is one of the goals.

The brain is the colony of mind, and any event in nature with any pattern of elements, in any complexity, when entered as an input sensory, provides a multidimensional sensual map of activated neuron networks. The map would be resonated for the memory of the elements as already saved, and those not recognized in this comparison will remain as stressful elements to be saved by the relaxation property of the brain.

All the time, new entries would be recognized by new patterns to be screened for memory available and to make a memory that is new to the brain. In the process of perception, interaction, responding, and making memory, the brain goes under permanent changes. Furthermore, the brain is not a perfect media, and therefore, perceptions and cognitions due to these changes are not prefect too.

Created patterns of an activated pathway is a dynamic bed of free energy transfer. The free energy transfer is the result of a stress flow initiated by an input and causes a current of synchronized firings to proceed along a pathway.

It is believed that such physical phenomena are the basis of different types of brain activities, depending on the type of pathways and the areas participated in the activity.

Each neuronal network in the brain is a substrate for a packet of energy during the free energy transfer. Because neurons in each location had grown in fractal shapes, it is expected that a firing cloud over any activated neuron network shapes a fractal too.

Attractors for the fractal have different categories, from basic structures in the inner brain (which arouse instincts when exited) to very complicated neural networks structures (which hosts complicated psychological desires). A pathway of synchronized firings develops seeking contingent attractors.

An energy flow consisting of synchronized firings, which is imposed by coming stresses, either increases the strains in new places or relieves that. In case that imposes more strains over the covering substrates, those strains would be absorbed up to the elasticity of the tissues on the location and would be overstrained in case of excessive stress. The overstrained location confines energy, which should be released to recover elasticity. This recovery happens in sleep periods.

What Makes the Brain Different?

The accidental changes in nature are limited to natural selected routes. There is only one possible route for action in one real time and one real location. The number of options for natural selected pathways in the brain is many. In fact, the natural selection is extraordinarily improved by creations of the brain. The ending product for a brain activity normally has different choices to select one action when such an advantage is not available in the nature by itself. The capability to search more options for actions (pathways) is because the brain can create a higher degree of abstraction, which provides a higher degree of freedom.

The mind is the result of such higher freedom of media, which can be defined as a cloud of patterns and synchronized firings in orbits and fractals, free from straight substance-to-substance and event-to-event links in nature. It floats by energy packets flowing in fluctuating pathways.

The brain is a colony for the mind by its substrates for energy packets, where each energy packet is a definite configuration of synchronized firings. Contingency between inputs, memories, and attractors provide faster and predictable endings, which are impossible by themselves in the nature beyond the brain.

However, the brain, with such an advantage, has limitations in the physical property of energy absorption-desorption because it can be strained and is limited only to its elasticity. The confinement of energy due to strains reduces its elasticity, which needs to be regenerated and recovered to previous condition.

The recovery of elasticity happens during sleep by the consolidation or release of the confined energy as memory building and dreaming processes.

More about Sensory Input and Stress

An input entry is evaluated for qualification regarding its need satisfaction and induces a feeling of pleasure if satisfactory and pain if unsatisfactory. The evaluation for quality regarding the need satisfaction is interpreted as a circumstance. Satisfaction of an unmet need as the *source* of input stress and consequent strain on neural networks will create a positive emotion (fulfilling the unmet needs) as well as negative emotion (highlighting the unmet needs) according to the circumstance.

According to Maslow's theory, "People are motivated to satisfy unmet needs," including physical, safety, social, esteem, and self-actualization. When a more basic need is satisfied, then a *higher level of need would emerge in the brain to be satisfied*. All needs are attractors for the pathway a brain activity finds. The resulting images and thoughts in the brain should initiate those motivations.

Basic needs are more inner-brain oriented, while advanced ones are more cortex and frontal lobe oriented. Social, esteem, and self-actualization needs are especially associated with self-prestige. It is highly stressful when self-prestige needs are attacked.

If any of these needs, instead of being satisfied, would be ignored or prevented or attacked even by oneself, it imposes high stress patterns over the brain. The overall hierarchy of importance is the same as above, but they may make complex needs an attractor to direct the way images would be created.

The simulation starts in different pathways of resonating memories and recalling inputs. When it is less satisfactory, the excited pathway would find harder stability. Furthermore, the farther from stability, the stronger the induced strains will be. There are more advanced areas like the frontal lobe, which facilitates the stability of the pathway if it would be involved in the brain activity.

Strain-stress effects in course of a brain activity

The firing particles if synchronized within a substrate are continuums in a brain activity, and they are to make neural networks connectivity. The force which binds the neuron and neural networks together in a pathway for brain activity is in fact merely a remnant of the strong interaction which the fine electromagnetic forces impose. The strong interaction is mediated by firings, and the force with which they bind the synapses and networks together is so strong that they remain concentrated on one dominant pathway, except at the highest strains points on the layer where it start to deform and firings become irregularzed.

The only known way to study the kinematics explanation of the brain activity without approximations in the tissues stress-strain behaviour and changes on the firing gates sizes and orientation is to place space-time on a four dimensional lattice, so that the theory can be simulated on a computer. This kind of the lattice simulation, may need computers with a very high performance.

Of the four basic forces of nature including the strong nuclear force, the electromagnetic force, the weak nuclear force and gravity, the electromagnetic force is the one applicable to the brain activities.

In addition to electric charge, firings possess a type of strength called interval time shortness. Firings with high strength experience the strong interaction, which is mediated by the static electromagnetic field. The theory which describes the strong interactions by resonating between firings in a neural network and special incomings including attractor influences is called strong connectivity building, memory or specialty generation.

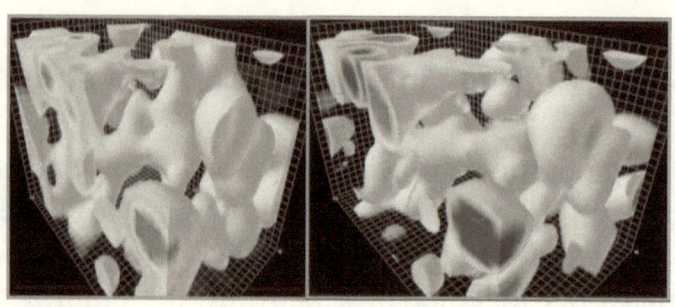

What is a Pathway?

A pathway is the bed stream for a regulated free energy flow by attractors and is made of substrates with a definite structure of neuron networks that build the pathway. Some substrates in the outer layer find a same structure with substrates in other parts of the brain during excitation with coming energy and resonate with them. The resonating far substrates are the attractors, which regulate the pathway.

Physical, safety, social, esteem, and self-actualization needs are physically definite configurations of neural networks, which are located in different locations in the brain, depending on their complexity, and act as attractors to direct the pathways' directions. The other type of neural configurations saves different types of memories that were made by past experiences, describing interactions of entry inputs and needs in different shapes and how much they were satisfactory or not.

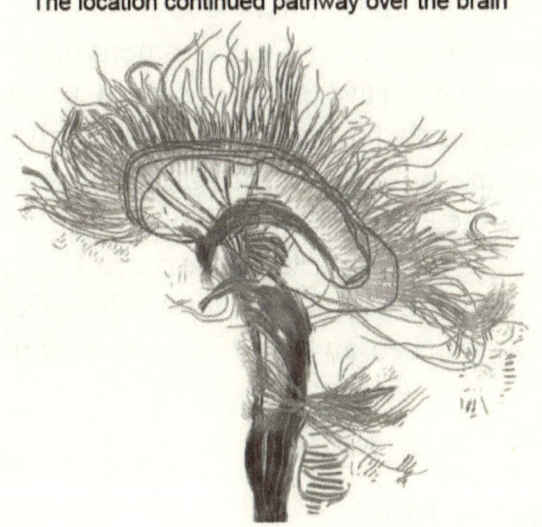

The location continued pathway over the brain

Circumstances and Emotions

A circumstance is defined as an interaction of an input in regard to a need satisfaction when the previous memories of circumstances are encoded as emotions.

A circumstance challenges how much it can satisfy a need by protection from attack, providing or promising, and corresponding to the degree of satisfaction; it is stressful. Circumstances evoke emotion, and evoked emotion decomposes to a variety of possible images of memories when released energy from a strained location lets it go through other neural networks while the release follows the elasticity of related tissues. A real circumstance initiates an emotion by entering input; an emotion decomposes into linked memories in the way the pathway conveys. A circumstance acknowledgement can be the perception of a real sensory input or an image aroused from a strained location.

In sleep, there is no sensory input. Therefore, a dream is a virtual circumstance that is in a strained location because of a stressful previous input that had been made.

Real circumstances in the environment do not care about our needs, but they leave emotions with us, which are all about those careless behaviors of need satisfaction. Therefore, circumstances are not meaningful on their own but are very meaningful by the emotions they arouse. The daily emotions reinforced during waking releases a stress flow through next neural networks to the strained location, which fresh imaginary scenarios created based on stressful narratives popping up as dreams.

Creations in the Mind

Declarative and nondeclarative memories saved in neural networks would resonate with neural networks in the outer layer whenever their structures permit physically according to the resonating rules. Because of a wider surface and type of neural networks, the elasticity in the outer layer is far more than in the middle brain and inner brain, which should have a more rigid structure.

The outer layer is much more flexible in elasticity limit. During the configuration fluctuation of networks, a network can find a similar configuration from another network in other regions and brain layers. In such a moment, those two networks resonate together. The intermediate structure of deformed networks during exciting links different networks in the moment, creating an imaginary scenario that hardly happens in nature. This combination is the process of a brain activity, which is very abstract, and nature—without brain—lacks that. The combination of momentary changes in configurations of neuron fibers, synapses, and regulated firings instead of random firings in rest is an extraordinary phenomena. The strain in the new configuration confines an energy that evokes a related emotion. The accumulated strains reduce the elasticity of the outer layer and are needed to be recovered by releasing. Relief from strain is done by releasing the confined energy around the strained location, which excites a covered neural networks around presenting a dream.

In short, the main topics from the sections in previous volumes of the book are listed as the following:

- Stressing and relieving of different pathways (neuron connectivity) is one of the main functions of the brain.
- A pathway is a stream bed for firing energy transfers directed by resonating locations in the course of one unique mentation. A pathway-strained pattern is nonstructured and cloudy when initiated but develops to a clear wake-like stream with definite fractal structures in parts. If the natural frequency of this terminating state can be graphed, a plan of neural networks' amplitudes and frequencies would be created with surfaces, hills, and valleys. The stabled points of the graph-like valleys indicate objects or characters, and unstable parts of hills indicate behaviors of the subjects, and the whole map shows the dynamics of the scenario, which conveys the meaning of the brain activity.
- The brain is a viscoelastic material with a time-dependent relationship between stress and strain.
- During the day, the layer becomes imbalanced in strain distribution.

- Some areas approach their highest range of stress absorption because of the accumulation of the confined strains. The pathways through those areas become cloudier and not determinative.
- Attractors, which direct the pathways' progress, are our self's needs and desires of different categories when self-protection and self-advancement are the core for all those attractions.
- Periodical regeneration is required to recover the viscoelasticity of the layer due to daily accumulating strains and the degradation of elasticity.
- Any category of inputs simulates different attractors in different locations of the brain with firing fractal resonations. The inputs are processed in the outer layer with higher elasticity and stored in the middle layer with its lower elastic properties.
- The stress release of the strained locations is attached to emotion content, which is a safety alarm for overstrains in different neuron-fiber configurations.
- Strains are a result of self-unmet needs and wants or stressful self-oriented events, which increase energy transfers in related pathways. Consequently, during release in the regeneration period, they reflect the same emotion as was felt during the related stressful input because the same configuration is stimulated.
- Overstrained locations generally undergo an instant reconfiguration and a linear releasing process afterward. This process can happen in stages. Therefore, after releasing the bundle of fibers in a pathway, the pathway generally returns to origin positions except for a residue of energy, which remains but will be used as fine reconfiguration of synapses, saving the cause and process of the related stressful event.

- There are two directions of energy transfer: by the growing of firing fractals from the synapse level to the layer level (through different levels of neural network resolutions) and reverse, when energy transfer is terminated and the built energy in the pathway transfers the energy to motor neurons.
- During sleep, the initiation is from strained locations on the overall layer, submitting energy to produce firing fractals in the synapse level to restrain the layer thoroughly. While a pathway during a brain activity is complex and lengthy, generally in a longitudinal direction and with distributed strains, it is spatial through neural networks from lower resolution levels to a higher resolution layer and with concentrated strains. Therefore, a dream is much more emotional than any daily brain activity.

In this volume, these headlines will be integrated to build a logical basis to develop a model for brain activity modeling to enable us to trace stressful events on a daily basis and induce emotion release in dreams during sleep, knowing a previous memory that contents the same emotion. A software package which aims to utilize all these principles to predict the emotion in a dream and somehow present a possible dream, which I hope this will be developed in the near future.

Applications of the Dream Analysis

One of applications of the dream analysis is facilitating to identify, explore, and manage one's emotional experiences, which is the key element for example in emotionally focused therapy or emotional freedom technique.

A general definition for emotion is the following: "Emotions are physiological neuroendocrine responses to which we react, when they come into awareness, with thoughts and feelings about those feelings."[1]

In the kinematics model, emotions are the excitation of overstrained locations in the brain that had been experienced sometime before in life and will be experienced similarly by a daily stressful event, which causes a similar type of overstrain in a location again.

Emotion, as the happening of overstrains, can be attached to any mentation. Practically any saved memory has happened in the course of a strain release. Depending on the strength of memories that are used in a mentation, emotion can be felt more or less intensively. It is strong during automatic thoughts but moderate for superconscious activities. In this respect, emotion accompanies the response of the neuron tissues to an overstrain mark point to retain a safe position, either during experiencing the overstress or recalling the memory. The higher the tension due to higher stressful events, the higher the emotion intensity and the stronger the drive to remind the patterns that had experienced the same tensions.

By time, an emotion's intensity will change because of new strain impacts in the location and to reconfigure the previous type of synapses and the fibers' positions.

In its extremes, because of a lasting or a shock type of stresses, non-normal strains can cause mental disorders by making those brain circuits solid by plasticizing some circuits further.

---

[1] Myers, David G.; Theories of Emotion; *2004*

Emotional health

An emotion history during the life should be an indication of mental health, because it shows if the brain regeneration had occurred as required and no strain accumulation created undesired connections in or between neural networks in different levels. The torturing of the prisoners by drops of water or other ways has been reported as the worst way to force prisoners confess either to true or false statements, so that they could not find a mental health afterwards.

> sleep restriction tends to make people less optimistic and less sociable subjects limited to four to five hours of sleep per night for one week reported feeling more stressed, angry and sad. Their moods improved dramatically when they resumed normal sleep.[2]

In opposite, when one is emotionally excited cannot go to sleep easily.

> emotional upset can severely impact sleep.
> sleep is clearly vital to emotional well-being

According to the kinematic model, in two different types of sleep stages two different processes of reconfigurations in synaptic level and distribution of concentrated confined energies to balance the layers of the brain occurs. There is no way at present to monitor the synapses reconfigurations and no knowledge to define a healthy new networks consolidation. But the dream monitoring for emotional content and studying

---

[2] Why Dreams Are Vital to Emotional Health; *Dr. Andrew Weil*; 2012

the emotion distribution during days and months, can provide a good index that how healthy is the mental life.

> mood disorders are strongly linked to abnormal patterns of dreaming.

> individuals who dream and remember their dreams heal more quickly from depressive moods associated with divorce.[3]

It is not known that lack of dream as reported by some people is a fact or because they cannot remember the dreams. However, interrupting the dreams is always followed by complain if it was a happy dream; and practically follows awaking if it was a nightmare.

> "dream loss" rather than sleep loss *per se*, is "the most critical overlooked socio-cultural force" in the development of depression.[4]

> many medications used to help people sleep also suppress dreaming.

> Many antidepressant drugs suppress dreaming as well.

The highest strains would be naturally released first, although it is not an always rule. Therefore, nightmares are expected to appear in beginning of the sleep. The statistical research of REM and NREM stages shows that NREM stages are longer and more frequent in beginning of the sleep; while,

---

[3] Why Dreams Are Vital to Emotional Health; *Dr. Andrew Weil*; 2012
[4] Why Dreams Are Vital to Emotional Health; *Dr. Andrew Weil*; 2012

REM stages are more frequent and lengthy in end of daily sleep.

> dreams "in the first part of the night appear to process and diffuse residual negative emotion from the waking day; dreams later in the night then integrate this material into one's sense of self."[5]

Brain Activity Parameters and Processes

The main variables in the brain are *substrates* (neural networks as the site for memories to recall or to save), *pathways* (with sub-variables of: attention, focus, speed, spatial reasoning, problem solving, delights, reaction time); and *expressions and actions command* (emotional or logical).

The above parameters from a kinematic point of view are defined as follows:

- Substrates: are the neural networks along a pathway which carry a firing energy packet accompanying a stress flow. They refresh or save a kind of memory whenever activated.
- Memory: as a solid configuration of synapses in a neural network is made during a stress flow, will be consolidated during sleep and will be recalled when the hosting neural network would be activated. The crystallized configuration is capable of creating a fractal shape of firings according to the firing gate spatial distribution to represent a copy of inputs.

---

[5] Why Dreams Are Vital to Emotional Health; *Dr. Andrew Weil;* 2012

- Pathway as already defined, is the bed stream for a regulated free energy and is defined by parameters like attracting needs, attention, speed, reaction time, direction and destination (delights, problem solving, spatial reasoning). A brief definition of these parameters are as follows:

    1. Attention: An advanced attractor that discriminates the inputs by keeping a definite pathway of a working memory alive.
    2. Speed: Determines how fast a brain activity takes from the initiation to termination. The speed is determined by the reaction times in consisting substrates.
    3. Reaction time: The time from initiation of a brain activity to its termination, which is an integration of consisting substrate reaction times through the pathway.
    4. Spatial reasoning: The process of the geographical coordination determination for a definite input entry
    5. Problem solving: The continuation process of a mentation during a complex brain activity searching an output to satisfy a need
    6. Delights: The result that a brain activity ends in without attention involvement

- *expressions and actions command* are a brain activity output which has the middle and frontal lobe involved in the pathway.

Needs and Attractors

Human motivation is for need satisfaction, and the following motives confirm the general brain patterns of neuron connectivity and configuration:

1. The basic *physiological* needs, which are required for surviving (like food, water, shelter, and sex satisfaction), to be met
2. Being secure or seeking *security* at home and in society
3. Being or becoming *recognized* and acknowledged by others
4. Being or becoming *respected*
5. Being *able to develop* to fullest capacity

The first three needs are primary and, if not met daily, create strong strains as negative emotions; and they if continue to remain unmet, they establish an unhappy mood. Any of these needs if is satisfied, the next level of the need become a drive or an attractor for motivation. The first three circumstances are essential for surviving and, if not satisfied, build a strong negative emotion. However, they mostly do not work as positive emotional drives. But the last two circumstances are drives and attractors for both positive and negative emotions.

A challenge of daily sensory inputs and these circumstances, either physically or psychologically, determine the brain's overstrained locations and attractors for the same night's dreams.

In short, the brain communicates by media of patterns with itself or its surroundings. In other words, by inputting through sensory channels, physical properties convert into the stress flow patterns, and by outputting, strain patterns convert into equities, and their symbol is saved in the input-output data bank.

The following terminology will be used in dream image presentations:

1. *Loyalty* is the degree of the conflict and intensity of the overstrains or emotions.
2. *Emotion awareness* is how quickly the associated images rise to a strained structure in the head or how quickly the related saved patterns (memories) are retrieved.
3. *Patterns* are the synapses' spatial configurations, which are the platform for related synchronized firing fractals in case of excitation.
4. *Pattern equity* is a combination of pattern awareness, loyalty, perceived quality, images, and emotion. Equity here is replaced for properties, which are associated and attached to a substance, event, or action in the environment.

A highly stressful pattern emerges from related emotions and attached firing fractals and images when it is being released during brain regeneration (consolidation of fibers or synapse configuration and uniformity of configuration energy over the layers).

Any stress-release pattern follows a pathway, which is bounded in boundaries. It functions accompanying changes in firing gate distributions and their geometry. Changing spatial distribution creates changing firing fractals, which seem like flowing energy packets in the pathway and a presentation of images of dream scenes and motions in a narrative. It means the pathway guides the narrative. In frequent scenarios, social and esteem issues are the most stressful sources to initiate the dreams.

The sequence of dreams at night is roughly in an order from the highest-strained pathways to the lowest, and sometimes there is a chaotic scenario cored by a complex of pathways. The most stressful pathways will be released earlier after sleeping, sometimes as nightmares, and the ultimate source of dreams is the self-actualization of suppressed dreams.

Unmet needs, if achievable, create achievers and, if attacked and destroyed, would create achieving dreamers (for self-actualized characters) or challenging dreamers (for passive characters) with a positive or negative feeling as it was felt during the attack.

Desires are a kind of advanced needs that are wished to be satisfied, although are brain-made needs and less available, therefore, a brain activity is required to bring them in the needs' level. This is an essential creativity of the brain's activities in humans, which are distinguished from brain activities in other animals.

Copy images of external or internal-oriented inputs are perceptional images of needs that already exist outside, but *creative images* of desires are new images of things to be created and are only internal oriented inputs. The state of consciousness about inputs is associated to having images as either copy images or creative images. The *state of superconsciousness* is the state of having creative images dominantly. Esteem and self-advancement initiators are kinds of desires that do not seemed to be looked for by animals.

Stress flow Direction

Strains initially increase in the outer layer where inputs are entering. Following that, other layers absorb the strains for a temporary balance in energy distribution. By resonating with different neural networks in different areas, strains will develop in the middle layer, frontal lobe, and inner brain.

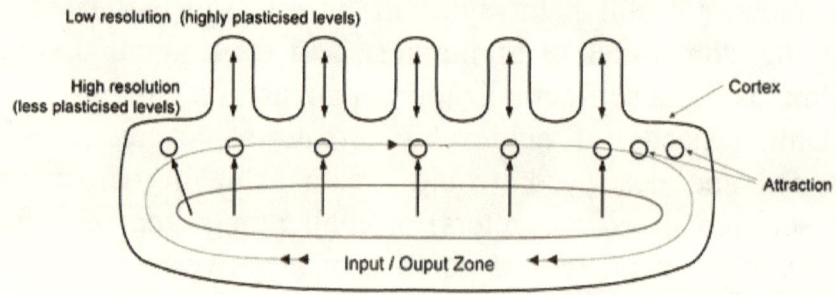

It is assumed that the appearing and disappearing firing fractals are associated to the *brain activities* and are made on the basis of general human brain structure and its topological curvatures, the layers' and tissues' stress-strain characteristics, synapses growths, and periodical reconfigurations marked by stressful events or trainings, and the patterns saved in the molecular and cellular scale.

In a kinematical view, there is a state of equilibrium in the rest condition of a layer topology and a state of understressed condition in the layer geometry and a continuous fluctuation of the later conditions around the former.

Individual stressful events in life make the majority of the changes in the layers' curvatures, and it is important in building one's character.

During sleep, the direction of energy flow is as follows:

- Location/s in the higher level of the neural network resolution of the outer layer would be highly strained by daily stresses. This location/s during strain releasing create the firing fractals.
- The relationship between the firing fractals and images are not known; however, the fractals are at least a

representative of them. The strains required to change the firing gate distributions from a random firing to a synchronized firing condition induce the emotion and feeling theme attached to the condition.
- The declarative and nondeclarative memories, which convey the same emotion and feeling, produce the narrative based on that combination and create scenarios.
- The above steps can be summarized as: strain build ups, feeling functions, and memory consolidating and visualizing.

The direction of energy flow in the release process is from a strained level of a neural network distributing spatially over the layer.

Therefore, opposite to the waking type of brain function, there is a sequence of stress distribution from the overall layer toward synapse level, resonating with neural networks in other regions and developing firing fractals by the stress distribution.

Any strained location distributes its confined energy to higher levels of neural networks because of higher elasticity, transferring changes down to synapses. Due to the chaotic condition of the media, the *characters* are not determinative, but *one of the same family characters* which convey strongly the emotion core—and is hosted in one of the neural networks in that strained location—recall other characters which convey the same emotion. In this way, a family of networks convey a feeling close to one another.

The sequence of dreams most probably starts with locations, which are strained higher and contain a higher amount of confined energy to release. However, this depends on the required activation energy of the neural network in that location as well. In general, it is not very deterministic too.

Those neural networks that host memories with close feelings represent metaphorical and simile symbols as well. Any symbol, due to its memory assignment to a network structure, is associated to a combined frequency bundle: a natural frequency. A dream is a collection of those connections from one unique strained location. A dream may be followed by a consolidation of a network that had contained a high confined energy and had released during dreaming. The synapse level is the level that communicates with other neural networks in other areas, including areas in the middle brain, through evoking firing fractals and resonating.

Connectivity in the lower-level portion of neural networks is more rigid when the connection in the higher degree of resolutions is more loose.

Free energy transfer along layers occur through higher-resolution network levels, especially the synapse level in a pathway, and interactions with substrates through the pathway happen spatially with the lower networks' resolution levels.

Sleep is the relaxation process strains in a location to uniform strains between different resolution levels of neural networks in the layer spatially.

Defining the Free Energy Flow Attached to a Brain Activity

Firings, chemical changes, and electromagnetic field fluctuations (in molecular, cellular, and fine neural networks) and substrate, stress flow, and geometry changes (in lower resolutions of neural networks) are parameters that play a main role in the brain to run a brain activity attached to a free energy flow.

On the other side, if we were able to monitor all changes in the neural networks of lower-level resolutions in a course of milliseconds and by location, there would be a dynamic flow of energy eddies, which would cause some changes in synapse curvatures, chemicals, and firing properties.

If we look at folds as rigid sharp discontinuities in stress flow, neural networks with a medium resolution as a semielastic media for stress flow, and straining eddies as the energy lumps, then any initiation by sensory or internal inputs will motivate a free energy flow made of energy eddies utilizing the above phenomena. The free energy flow terminates in the output zone. The standing configuration of the flow pathway after termination is a performed brain activity. At the beginning of a clear brain activity, energy flow has a random velocity and fluctuations around in different directions. It is turbulent and chaotic. Due to the low thickness of the layers, the flow in vertical direction would be damped, and finally, when energy eddies are absorbed through the pathway, the brain activity is ended and ready to dispatch the releasing energy to motor outputs.

Energy eddies are continuously and constantly interacting together (with negligible time lags) so that they cannot be recognized as individual units with present measuring devices.

The eddy flow in the outer layer's backside has a wide range of sizes and therefore is most chaotic. The size of eddies is proportional to the square of their distances from big plasticized curvatures in lower-resolution neural networks and finally folds. In the continuation of their transfer, they move in a more steady state condition, and finally they find a creeping shape of a flow of eddies configured in a complicated shape as the product.

Characteristic length as an indication of eddy size is proportional to distance of the eddy from the folds.

MOSTAFA M. DINI

Stress Flow in Calculation Terms

Although this section may seems not similar to other sections in the way of description, it is added here to introduce kinematic terms, which will be used in calculation stages later.

Eddy viscosity, which is a strong function of position and time, is a product of *energy density* and *stress diffusivity* in the pathway. *Stress diffusivity* is a product of the substrate section (the network host of the eddy) and a second-order derivative of the stress-rate function along the layer. Therefore, eddy viscosity is very high in the backside while finally negligible in the outer layer's front side.

The free energy flow is as important as the required activation energy in locations along the pathway.

A distribution of the required activation energy determines the direction of the pathway.

Big eddies hosted in lower-resolution neural networks appear as form drags and move very regularly as definite orbits. When *eddies* by interactions in a pathway find a bigger size, they form more static drags, and because of higher inertia, they tend to behave less energetically. A mass of eddies shape a wake of energy with low-traveling velocity. The streamlines in a wake of energy are very regularized and creep in laminar lines.

Drags are a locally and timely distribution of the individual eddies and have a dynamic pattern with a spatial rotating shape of eddies. Drags move forward in the pathway and make wake patterns. Stress convection flux due to eddies, drags, and wakes, in different directions, is a product of related mean stream flux differentials and energy density. It is also the negative product of eddy viscosity and the second-order differential of the stream flux along the layer direction, while viscosity itself is a product of density and eddy diffusion. Therefore, it can be concluded that eddy diffusion is a product of the eddy's substrate surface and the stream function's second derivative.

Conceptually, if eddy configurations are considered as the alphabet, *drags* make words and words build *wakes* of sentences and all together can be translated as a brain activity statement.

The specialty of an area in the brain, to some extent, shows how the flow streams are shaped during a stress flow through that area. It indicates how much important the neural network types and configurations are on the location. The finer the energy particles, the faster they move. The chance of larger sizes is more to be absorbed for the energy, and by that, the resulted confined stress would increase the required activation energy of the location. This enables the location to possibly initiate an internal input when it reaches to dispatch its confined energy. Furthermore, when the location is under process of energy absorption or desorption, its amplitude of oscillation for an absorption-desorption task is stronger.

It can also be concluded that the increase of branches (jumping from a subject to another subject) in a brain activity is limited in the frontal lobe (by its neural networks of higher-level resolutions) and therefore can more regularize the stress flows through it rather than other parts of the cortex. In other words, higher-level resolution networks are finer networks that can accommodate finer energy packets and show high elasticity. The *regularized streams* can physically describe the *logic functions or reasoning*.

In short, the flow of stress throughout the cortex finds a chaotic, transmittal, laminar, and finally structured shape of the streamlines, which somehow represents initial perception, cognition, reasoning, and conclusion in a course of brain activity.

MOSTAFA M. DINI

Resonations Between Brain Regions

Pathways in the outer layer are considered as the main route for communication between neurons in different regions and areas because of higher surface and connections to input-output zones. In addition to it, connectivity for stress flow is highest because of higher elasticity. However, any pathway and its direction in the outer layer are determined by the attractors in other brain layers, which resonate with networks in the pathway.

Excerpts from experts may help to confirm the above:

> Areas that are physically close have more connections with each other than they do with parts that are farther away. But . . . a number of regions collect information and route it to more distant destinations . . . . precuneus is particularly well connected to distant regions.[6]

> Over the last few years, functional neuroimaging studies have started unravelling unexpected functional attributes for the precuneus, a cortical region located in the posteromedial portion of the parietal lobe which has widespread connections with both associative areas and subcortical structures. The posteromedial parietal areas are amongst the brain structures displaying the highest resting metabolic rates ("hot spots") and are characterized by transient decreases in the tonic activity during engagement in non self-referential goal-directed tasks ("default

---

[6] Anne McIlroy, "Computing Methods Used to Study 'Servers' in Human Brain" *The Globe and Mail*, June 2011

mode of brain function"). On the other hand, precuneus activation has been documented in healthy subjects engaged in self-related mental representation and episodic/autobiographical memory retrieval. Moreover, selective precuneal hypometabolism has been reported in a wide range of altered conscious states, such as sleep, hypnosis, drug-induced anesthesia, vegetative state, and in neuropsychiatric conditions characterized by impaired consciousness (eg, Alzheimer's disease, epilepsy, schizophrenia). Taken together, these findings provide strong, albeit preliminary, evidence that this richly connected multimodal associative area belongs to the neural network subserving awareness and producing a conscious self-percept, a process that possibly runs in the background (by default) during silent rest.[7]

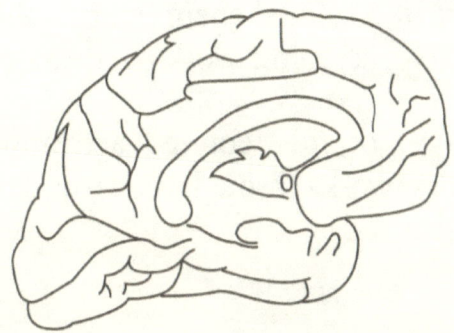

Adopted from Andrea E. Cavanna and Michael R. Trimble, *The Precuneus: A Review of its Functional Anatomy and Behavioural Correlates*

---

[7] Andrea E. Cavanna, MD, *CNS Spectrums*, 2007

"It is based on assessing the output of a region, the electric or magnetic signals it sends out."

"It is the timing of the signals that matters. If they are slow but steady, then not much is going on. But if the signals come all at once, then slow down, then speed up again, it is a sign of a lot of activity."

> "Variable output indicates that the processing in these brain regions is changing to meet the demands of incoming traffic."[8]

The Required Activation Energy of Substrates

The *required activation energy* for some areas is more accessible due to accumulated confined energy, and a fine increase of resonating energy on those locations makes energy levels to release energy for a more stable state. Such an elevated state of confined energy during the day keeps the related sites active, keeping working memory of the stress sources.

A combination of the required activation energies for different neural networks through the layer determines which pathway to be selected.

---

[8] Anne McILROY, Computing methods used to study 'servers' in human brain; The Globe and Mill, Monday, June, 2011

BRAIN FUNCTIONING AND REGENERATION

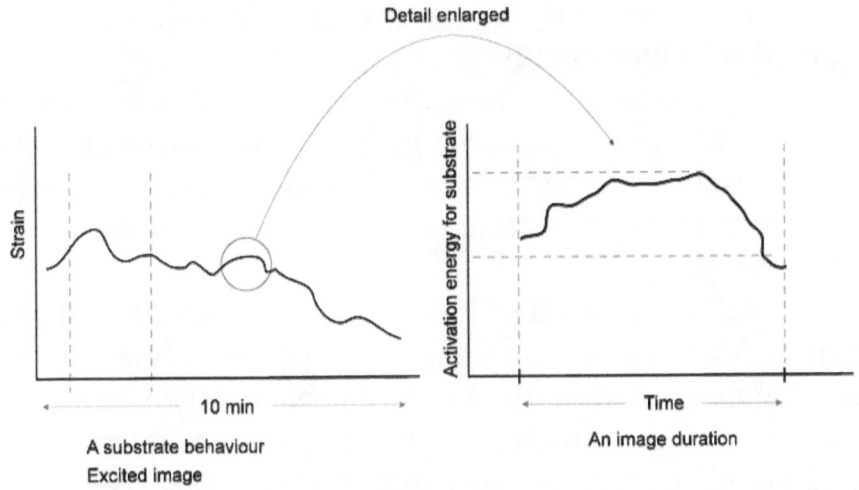

A substrate behaviour
Excited image

An image duration

A layer's curvature (hills and valleys) of connected neuron fibers indicates how input-free energy goes through the layer while the resonating and boosting energies on the way are considered. The free energy would be absorbed for a portion during travel as the confined energy, which makes the working memories and covers all memories saved in the elevated confined energy as strained locations. Physiological, social, safety, and satisfaction needs build such a pattern in relation to the degree of satisfaction of mostly internal stimuli while esteem and self-advancement needs form the pattern in relation to mostly external stimuli. All the stimuli satisfactions are due to comparisons between the needs' accessibility and accessibility expectation. The higher the gap, the higher the stressful inputs, which imposes higher strains.

The gap, as input information, would be processed in an excited pathway to achieve a decrease on the gap and reach higher satisfaction. Satisfaction is the target for any brain activity set as an immediate target or a target in the future.

In waking time, normal strains will be recovered frequently and only overstrains in a location remain as a confined energy due to no opportunity of discharging because of higher required activation energy.

In sleep, elevated confined areas, which keep the layer under strain, release the confined energy if it could be elevated to the location's required activation energy by support of a long-term memory network resonating; otherwise, regeneration would not be complete.

In NREM stages, those locations already with a confined energy, upon overcoming the required activation energy, reconfigure in synapse level to find a stable state geometrically. Such a reconfiguration changes the geography of the remaining area and will be followed by a new local stress release.

In REM stages, the local stress flow supported by the resonation of memories with the same pattern fulfill the required activation energy to start and goes through neighbor networks to equalize the location while resonating with networks of the long memory elements. This is the key concept to make metaphor and simile images appear. The appearance again is by synchronized firing fractals in the synapse level when straining forces configure the firing gates distribution and sizes in a definite way.

In conclusion, if emotions in our daily experiences are related to overstrained locations, releasing in dreams would refresh the same when they are reinforced by a past memory with the same emotional theme.

> *This means emotion themes are the same in stressful daily events experienced, supporting past memory patterns and dream content.*

Any fluctuation in confined energy content represents an image and a motion and is physically linked to a substrate in the releasing route. The number of fluctuations represents the number of images that play in a scenario of the dream.

Shadow images represent parallel branches of release routes. The total duration of a dream is a summation of the time frame of images (images and their motions), which makes a REM stage with an average of ten minutes. The strained substrates' average size (characteristic length) can be determined if a demonstration of energy changes in any location would be possible to measure in the future.

Consciousness decreases by the increase of the confined energy level to near its required activation energy and changes to a reflexive or unconscious level when it is above that level. It means the degree of consciousness is related to the amount of energy calling supporting memories intentionally to reach the required activation energy to overcome the specific route of neural networks. Such an intentional energy adding is possible by the recognition of the gap between achieving needs and available resources.

Stress Flow and Simultaneous Release of Strains

If strain changes would be drawn in a coordination system against the brain's outer layer, considering that the main pathway is assumed to happen over the cortex, *strains directly influence the outer layer and front lobe* and indirectly the middle and inner brains. The pattern of strain changes is similar to the Rayleigh surface wave which is a wave over rigid layers.

There is a *longitude transfer through the outer-layer neuron connections* and *front lobe* as well as the connectivity with the other parts of the brain. Accordingly, the locations, which may go under higher strains and sometimes overstrain, are outer layer—front lobe in first stage and middle brain and inner brain as secondary in sequence.

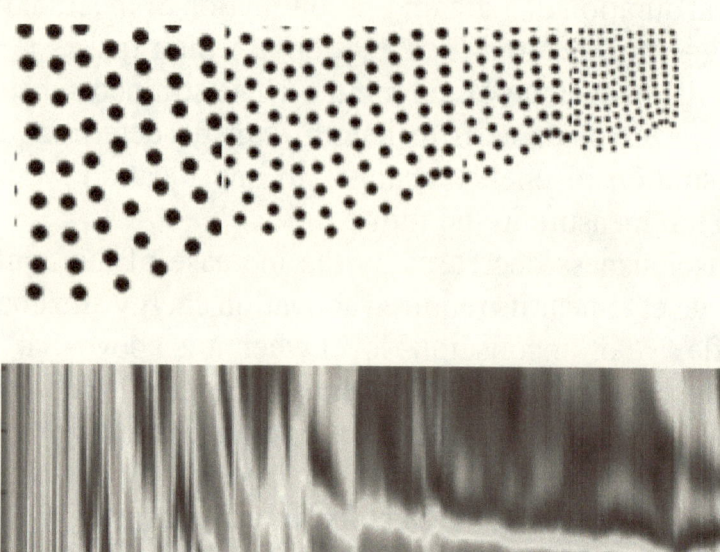

(Taken from Chris Barnes's works for conveying the Rayleigh waves in a different application.)

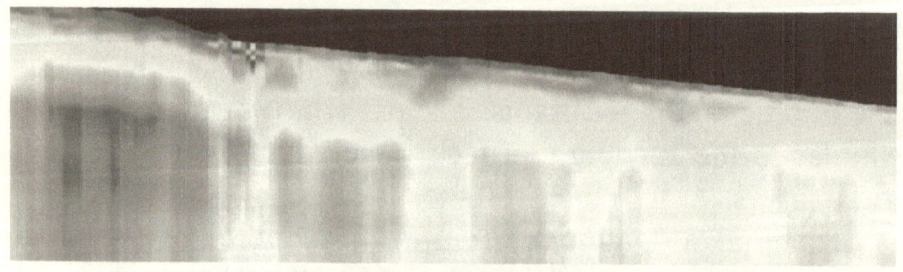

(Taken from WesternGeco Rayleigh Waves Inversion Noise Attenuation for a different application.)

BRAIN FUNCTIONING AND REGENERATION

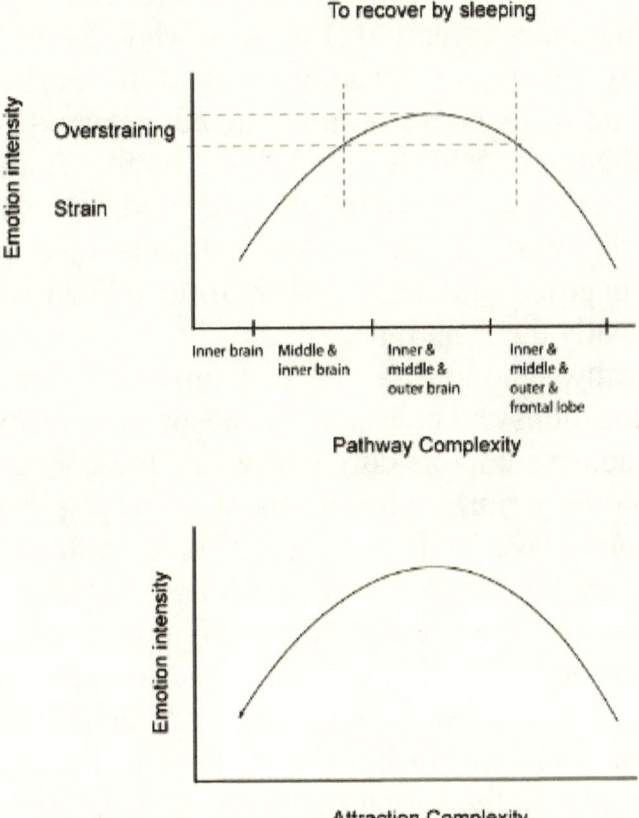

The outer layer and front lobe strains indirectly make the middle and inner brains strained as well.

Stressful events induce strains on the outer layer and front lobe. These layers strain the middle and inner brains, stimulating emotion and physiological needs.

As expected from the model, the highest accumulated strain in the beginning of sleep is absorbed by the middle brain during memory consolidations. It is assumed that the strain-releasing force influences the inner brain with actions like erection during sleep in males and eye moving as well. In higher strain-releasing cases, disorders like sleepwalking may occur.

REM dreams are mostly related to the recovery of wide areas of the outer layer and front lobe while NREM dreams are mostly related to consolidations of memories by the reconfiguration of high-resolution neural networks. Majority of consolidations are in the middle layer, with its fine structure ready to reconfigure. Therefore, the middle layer is the location for most declarative and nondeclarative memories while main procedural memories are formed in the outer layer and especially the front lobe.

Physically, emotion can be explained as overstrains on definite locations and consequently needs to be recovered.

Most active attractors during sleep are those located in the middle or inner brain, reinforcing the strain release in pathways over the outer layer and front lobe. Therefore, dreams utilize the active emotional attractors, which are located in different areas, especially the middle and inner brain.

Dreams with attractors within the middle brain are most emotional. Dreams with attractors in the front lobe mostly use logic and insights, and those in the inner brain are physiologically oriented like sexual, eye-moving, sound-hearing, or going-to-the-washroom dreams.

Similarly, romantic and dramatic dreams are initiated by attractors in the middle brain; dreams with socializing and esteem emotion aspects have attractors in the outer layer, and self-advancing dreams have attractors in the front lobe.

REM stages are full of scenarios composed of different location strain recoveries in the same moment. Therefore, a combination of different attractors can make the whole scenario complicated.

Strain-Stress Relationship and Absorption-Desorption Process

Sequential stress-strain traveling has a mix of longitudinal and vertical directions from the input to output's general direction. Therefore, the free energy transfers through area connections as well as the neural network from a higher-resolution level of synapses to eddy-energy packets in the medium lower-resolution level and to the lowest resolution networks of a layer.

On the opposite side, the layer tensions will be distributed by forwarding energy packet moves from lower-resolution networks to the medium levels and finally determine the synapses configurations in fine level to shape firing fractals.

A gross firing fractal in a location acts as an energy packet travelling through a layer from input to output. In symbolic language, synapses' level patterns join energy packets in the medium level of networks, finding a form of moves and sectional moves joining together in scale of the layer, creating a signal output. The energy transfer can die out in any section if the input energy is weak or the activation energy of the location is high or supporting memory resonances are not sufficient.

In the kinematic model, chaotic firing fractals in the synapse level within a substrate make changing configuration energy packets, and few of them together make a wake of energy. A diary memory element, which is supposed to be related to the most stressful event of the day, is analyzed for characters and behaviors. Then a fractal would be assigned to that according to the emotional symbols gathered in customized symbols, metaphors, and similes for the individual. In this way, images have gone under a personification process. The scenario would be a random combination of those symbolic characters and behaviors and background scenes. Indeed, past experiences welcome new ones to interpret them.

To design the software, (1) an assessment of strained pathways due to daily confined energies should be done; (2) the pathway emotion code is to be determined; (3) a cross-check in the data bank of past experiences with the same emotion code should find the matches; and (4) a personalized image of the same-emotion code is to be picked up if unique, or a random choice is to be performed if more than one, or related emotion intensity and type are to be considered in choosing if several similar experiences are found.

Strong *long-term experiences* in supporting the images to pick are those with the need-satisfaction index (physical, safety, social, esteem, and self-actualization) strongly not met (negative) or were met beyond expectation (positive), because they submit a stronger batch of resonating energy.

The strain caused by the energy wake on the layer needs to be released according to elasticity and will discharge the energy fractal, sending out to output motors and, in parallel, confining a portion to reconfigure synapse arrangement in the location as memory consolidation.

The free energy transfer is simultaneous and periodical, and it transfers in shape of helical moves through neuron connections. It remains travelling and activating the on-the-way networks up to full absorption of the energy along the layer. The activeness of a free energy pathway can be traced by working memory.

In conclusion, any strain-stress relationship study in any layer should consider the total neuron network connectivity; like a melt polymer plug, a strain-stress relationship considers chain-branching restrictions as well.

Experimental Stress-Strain Studies and the Kinematic Model

Assuming the stress flow is the physical medium for any brain activity, stress flow progresses and maintains a regular pattern of the firing gates' distribution and synchronization in firings until it is disturbed by the growth of a new pattern at least partially over the same pathway.

While partial relaxation after each periodical straining of neuron fibers and brain tissues cause to send out the output signals, strain-accumulating locations will go under a daily relaxation by (a) consolidation of fine deformations as memory and (b) releasing of the confined energy as the media for dreaming.

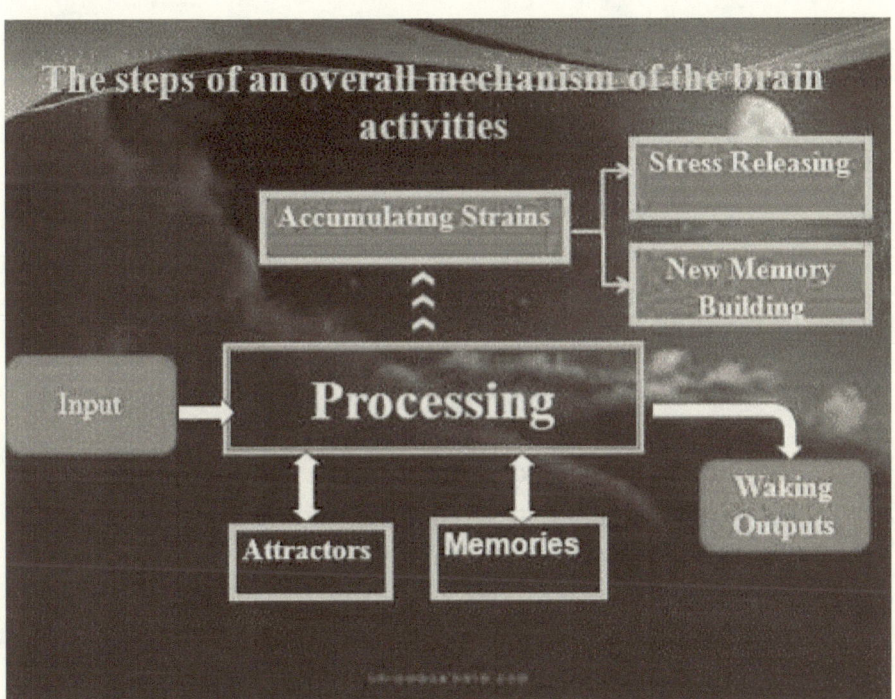

The experimental stress-strain studies confirm the interval energy outputting, consolidating, and releasing processes for the brain activities.

A summary of experimental data from the book *Neural Tissue Biomechanics* by Lynne E. Bilston shows the following:

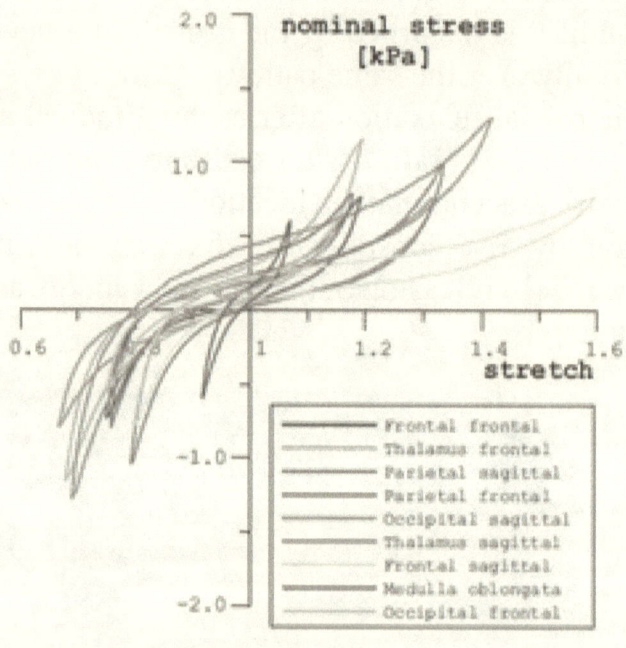

For the same range of strains,

> the outer layer and frontal lobe show a lower sensitivity to input stress, but the middle brain show a very high sensitivity to that;

> The middle brain covers a very wider range of strains, but the outer layer and the frontal lobe cover a smaller range.

The samples used in experiments are in strips. Therefore, it is expected that vital samples, which are made of a neural network of them, show more sensitivity to stress.

> The repeated loads cycles are performed after full relaxation, but a vital pathway will accumulate strains during the waking time because of connections in networks. Sleep is to recover the previous elastic property.

The tissue samples are taken from frontal and sagittal directions. An overstrained tissue shows a significant difference in two different direction strains for the same stress. Beyond a certain limit, the deformation will be *consolidate*d in a new configuration instead of the continuous appearance and disappearance of the temporary deformations. The relaxation is time dependent and retains the original neural networks' configuration, except newly added fine reconfigurations of the synapses' overstrained location.

The stress-strain data for tissues is as follows:

> There is a significant difference in stress-strain sensitivity in middle brain, than outer brain. In middle brain deformations consolidate for individual stresses; while outer brain needs a repetition of the same type of stress to consolidate a stress generation.

> Outer layer show less difference in strain responses in frontal and sagittal directions and facilitate a continuous stress absorbing-desorbing character, but the middle brain shows pulse stress absorption and desorption to save or recall a new configuration.

Higher cycles of load increase the tissues sensitivity to stress if they experience a full relaxation after removing the stress source. Vital tissues because of connections in networks do not experience a full relaxation during waking,

By accumulation of strains in the layer, it approaches a plastic limit and needs to recover for its elasticity property. In this recovery, *releasing* and *consolidating* of the temporary confined strains follow each other in stages of NREM and REM for the tissues with significant and moderate difference of stress-strain sensitivity in two directions in sequence. A computational study of stress-strain in future will be able to answer questions regarding brain efficiency and concentration capability in different times.

Free Energy Transfer Phenomena and Similar Functional Patterns in Different Circumstances

Emotions develop during a brain activity, with origin physical nature of pain and pleasure to high abstraction levels, while the main stream of the brain activity is a real free energy transfer phenomenon.

Different brain activities can have the same pattern if groups of neural networks involved in those activities introduce similar patterns of pathways in configuration. Those groups of networks may then represent one another as symbols.

Similar brain activity patterns are combinations of memories as scenes, characters, and behaviors or actions that indicate a circumstance for a need satisfaction with the same emotion.

If all types of emotions can be categorized as a tree, different ways have been recommended to do that. Furthermore, the basic emotions are introduced in different types. The types that are most compatible with the kinematic model are the basic emotions pain and pleasure. Pain is orientated from strains, and pleasure originates from releasing the next levels of emotions that are developed in the different branches of the tree. Each is a relay to retrieve specific types of circumstances in regard to a human need.

Each real circumstance has been experienced by a combination of substances and moves. Experiencing a real circumstance had also induced an emotion according the expectation of a need satisfaction from that circumstance.

A *psychological circumstance* can be defined as a combination of retrieved declarative memories and nondeclarative memories. Imagination of a virtual circumstance also brings up an emotion. In a tree categorization of emotions, there are emotions connected to one another in a local branch and emotions very distinguished from one another in branches apart. Therefore, different circumstances induce different emotions, but they can freshly close emotions.

An emotion is an encoding of physical strain combination in different neuron connections of a network. Therefore, strained networks with the same combination of strains recall the same emotion. Strains themselves are a function of stresses. In the kinematic model, the types of emotion and the emotion intensity are connected to the possible number of strain combinations in a network configuration and the position of the strains in a strain-stress coordination.

To apply the above concept to the brain, energy-interacting procedures are structured according to stress-strain rules in the outer brain; memories are saved in hypothalamus, and emotion codes are mapped in the amygdala and these are active during any brain activity, including dreaming.

During waking time, brain activities build some overstrained location in the outer layer, and the breakdown of the activities communicate with memories in the hippocampus for the same emotional pattern of memories; and during sleep, the overstrained locations in the outer brain release excess stresses to retain the original elasticity by consolidating new synapse configurations and distributing over neighbor neural networks by having communication with same-emotional memories in the hippocampus to evaluate the energy content by resonating to release. The same-emotional memory elements are relayed by emotion codes in the amygdale, letting it pick up related memories so that the combination of spread elements provides the same emotion. The consolidating process occurs by reconfiguration in synapses configuration. Both processes of consolidation and distribution of excess energy retain an equivalence state of distributed energy over the layer as an elastic material does during relaxation.

Daily Strain Changes and Resulting Firing Gate Distribution Changes

An example of the distribution of synapses and fibers in different layers of the brain:

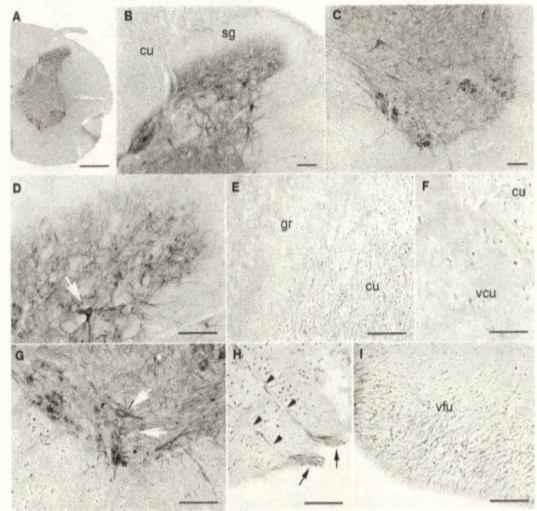

Fibers' in-out distribution between layers as well as brain-body:

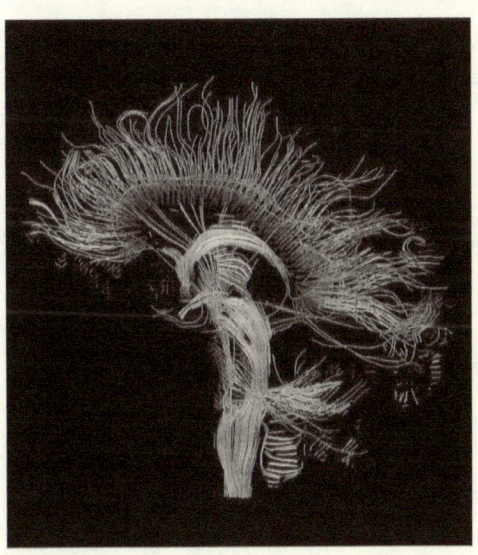

Samples of detail configuration of sample fibers in the brain:

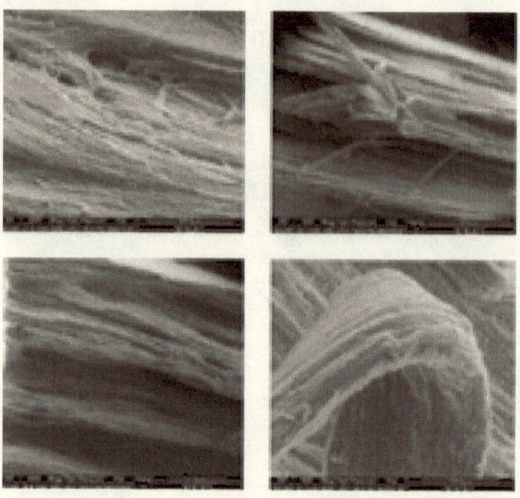

The *general trend of strain* increases with imposed stress on any strip tissue of the brain:

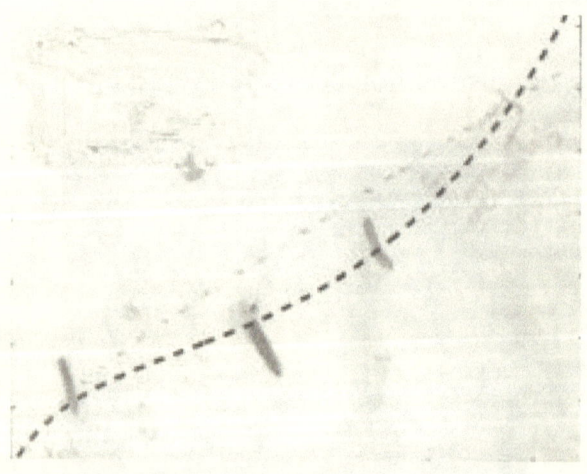

BRAIN FUNCTIONING AND REGENERATION

The general trend of increase in strains during a day:

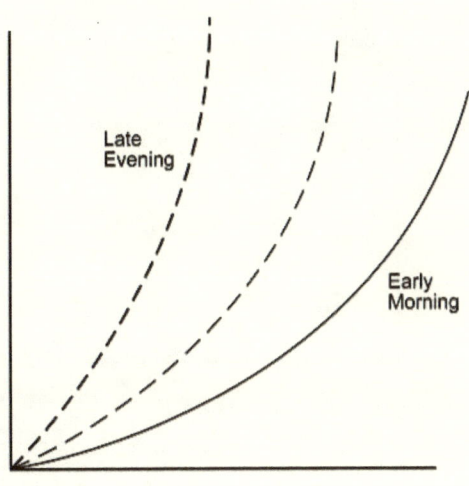

<42>

The *return of strained tissues during the regeneration period* will follow decreases in steps; however, again the general trend is similar to the increase of strains when exposed to stresses during waking time.

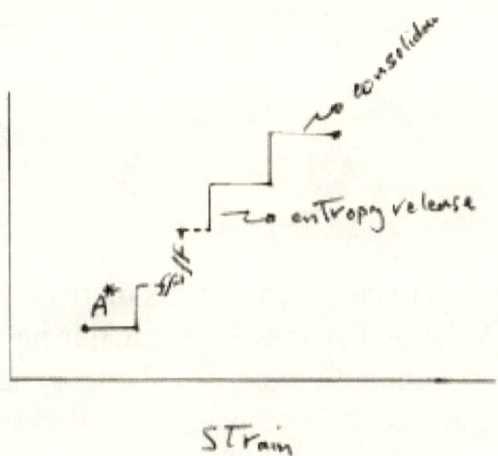

Top right is the strain condition in the beginning of sleep, and A* is the condition at the end of sleep.

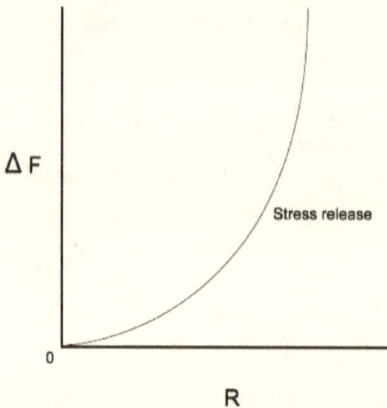

ΔF is the free energy released during regeneration (sleep).

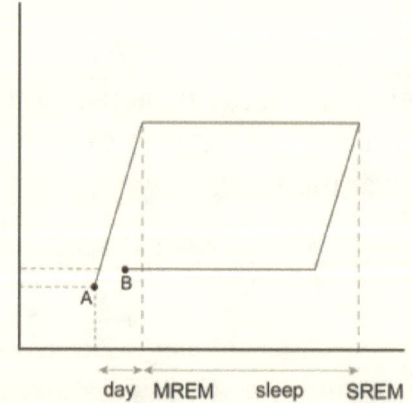

From point A up to *right left*, strains increase during a day, and from *top left* to point B, they decrease during regeneration (sleep). The lines indicate the confined energy-level changes during the day and sleep.

A typical complete cycle of increasing strain and the decreasing of it during the regeneration time is as follows:

The general trend changes based on the layer, region, and area:

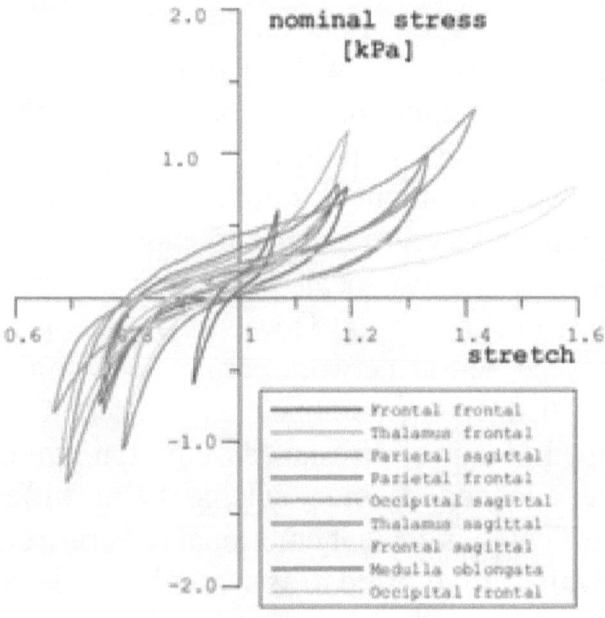

The trend of a pathway, considering its tissue bounds with surroundings, is in a spatial coordination, developing fractals in the synapse level to energy packets and energy waves through the layer:

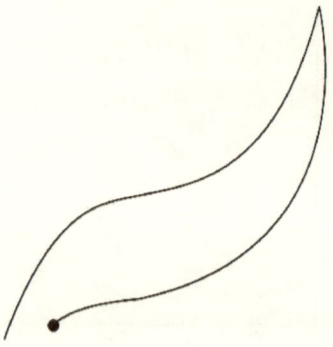

Growth and disappearance of the energy fractals is limited to the location geometrical synapses and fiber configurations.

In the course of any brain activity, a free energy flow would be developed during the waking time, with a loop pathway from input and output zones to the same locations by many branching feedbacks from different locations in the brain.

In opposite, the input or output sensory gates are closed during sleep, and there is no such loop possible. During sleep, the pathway is a short local connection of spread networks around the strained location. However, similar to day resonation processes, the neural networks around the strained location resonate with different memory sites in different locations. The overstrained location refreshes an emotion, which resonates the same encoded memory elements in different neural networks. These equivalent emotional patterns retrieve fractal images, which are referred to as symbols, and in combination, declarative and nondeclarative images present a scenario of dream.

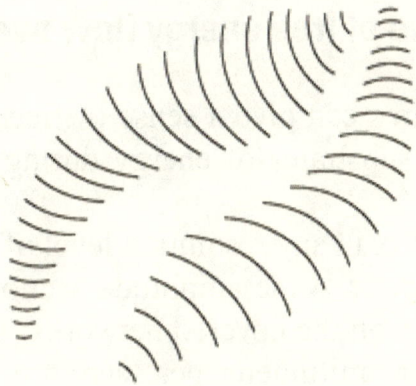

Energy Wakes Forms and Discharges through the Outer Layer

Each *firing fractal* makes one *energy packet*. The size and shape of energy packets change due to *fractal variations under influence of other resonating energy packets*. These variations seem like a *dynamic energy transfer* by having stress-strain effects over the pathway,

The mass of energy packets in a pathway together form a map of *energy in a wake shape*. The wake of energy would be discharged as a signal in motor outputs to relax the pathway tissues to relax according to elasticity rules.

The residue of nonreleased energies as *confined strains* releases from overstrained locations by *resonating with other areas of neural networks, including the middle-layer neural networks,* which increase their energy content and would find the equilibrium condition by *changes in some neuron connections as a new memory*. These changes happen during sleep.

Finding a sense of free energy flow over the layers

Flowing figures can give a sense of free energy amounts and the size of substrates of energy during the free energy flow.

The energy flux in any resolution level of neural networks is related to a power ¾ the amplitude sectional average of a transferring wave on the covered networks. Therefore, if it is 0.1 erg per square millimeter per second in a location, it is $0.1^{3/4} = 0.177828$ on the medium level and $0.177828^{3/4} = 0.273842$ on the overall layer. It could be in a range of a few to 100 microns in the lower level, up to 200 microns in medium, and a few hundred to two millimeters in the overall layer.

By assumption, declarative memories have been saved vertically over neural networks: component memories in the higher-resolution level of networks, intermediate memories composed of some characters and events in the medium-resolution level of networks, and those that cover larger memories cover a wide spectrum of resolution levels overall in the layer. By covering a wider range of neural network resolutions, the memory is more dynamic and more event descriptive rather than solid-facts descriptive. This fact should be considered in kinematic parameters of the free energy transfer through the layers.

Measuring Energy Rates of Free Energy Transfers along the Layers

One of possible ways to measure free energy flow is by tracing *oxygen content* of veinlets feeding and returning, which already is a regular procedure. However, the intervals and the procedure should be determined according to the goal.

The basis for the suggestion is brain nutrition is related to burning calories, and nutrition emery is the main carrier energy of the small energy flow between neuron fibers. The fact is, a *very high percentage of nutrition is consumed in the brain during sleep*. Having or approximating those percentages enable the conclusion of energy transfer rates by the implementation of oxygen takings.

Input decomposition into static, dynamic and circumstantial evaluation

A combination of character, behavior and the setting memories in hippocampus provide a scenario which is directed by an emotion in amygdala, as another part of middle brain;

If emotions are connected through a classifying emotion tree (in amygdala structure), then the memories are structured and possibly recalled in the same manner as the emotion tree is organized. The reversed phenomena of stimulating memory sites of characters, behavior and the setting demonstrate a possible dream;

Fluid blending of the participating characters, behaviors, settings according to an associated emotion demonstrate the process of the free energy flow in the outer brain;

Input sensory decompose in static and dynamic elements and the related circumstance against the needs and desires. During the processing stage them react, deform and combine according to neural networks which save them in specific configurations. The change in the configurations is dominantly under influence of new memory savings by aging. Similarly, the emotion memory—which is a circumstantial memory against satisfaction of needs and desires—evolutes and changes with aging. Emotion is selective in enhancing other related memories by associated dimensions of static, dynamic inputs. Other than sensory inputs, internal inputs also are combined of different static, dynamic and circumstantial evaluation dimensions. These types of inputs also decompose and integrate differently in course of processing. Specially the emotional type of memory has a dominant effect on the shape of the configuration sites for static and dynamic types. Therefore, a creation of a unique input memory consisting of its static, dynamic and emotional dimensions may be distorted compare to initial perception significantly.

> The sleeping brain seems to somehow make calculations about what to remember and what to forget.
>
> Sleep also helps to soothe the sharp edges of bad memories . . . . during rapid eye movements, or "dream" sleep, the hippocampus and amygdala reactivate, yet some arousal—inducing stress hormones, particularly noradrenaline, are suppressed. The lack of those stress hormones may let the brain process emotional memories in what seems like a safe environment. During slumber, he theorizes, the brain strengthens its memory of the information within a distressing episode while "stripping away the emotional tone."[9]

Amygdala is a memory place like hippocampus and other areas. This site save the aspect of circumstances dimension

---

[9] A Feeling for the Past, Emotion engraves the brain with vivid recollections but cleverly distorts your brain's record of what really took place; Ingfei Chen; Scientific American Mind; January/February 2012

of inputs and the overstrains they caused. During the dreams, relieve of the overstrained locations in the outer layer stimulates relevant site in the amygdala and the aroused emotion recalls any potential static and dynamic associated memory; which together present an imagination of a dream scenario.

Amygdala should have a capability to save a evoluted and organized structure to save a emotion tree like below.[10]

Building Up a Sense, Developing a Common Sense

Similar daily experiences stimulating specific neural network configurations would be plasticized in specific locations. The configuration expands to a family of neural networks of different resolutions in the same location because of the same neuron types in almost the same area. Consequently, the networks with the same required activation energy are *similar and will especially build up in the same location*. Activation of the same family neural networks should evoke similar feeling and emotions. As an example, memories with the same emotions pattern recall each other sequencely. This repetition can be encoded as the emotion of "feeling others as probable enemies in later contacts" or "feeling strange in groups" and would be saved as a common emotion for the self, and this emotion will intensify with similar experiences, and if it would not be controlled consciously, it generates an interpolated belief. The same family memories deform the neural network configurations in the same location and build

---

[10] A Feeling for the Past, Emotion engraves the brain with vivid recollections but cleverly distorts your brain's record of what really took place; Ingfei Chen; Scientific American Mind; January/February 2012

one unique emotion, which may be extrapolated for the similar experiences.

Because the overall properties of the brain are similar for different people, same types of emotions will be built during brain developing ages, especially if similar experiences had been observed. By building up of the same emotion with shared experiences, the common sense will develop between different individuals and make the communication possible.

Extroversionally, actions, behaviors, and expressions of common emotion will be performed by a *similar way of body language*, sound, smell, touch, or taste backed up in different individuals. It can be said that the background of those actions is the similar patterns of firing fractals, which had been created to motivate those extroverted actions. The *media for those extroverted actions are signals in the shape of symbols, which are formed in the prelanguage part of output zone and are more personal than common between individuals*. These are mostly pictorial symbols and more or less are exactly as experienced in the past without any development to more advanced symbols. One symbol will be used for the expression of a limited number of events with the same family pattern. These kinds of symbols are seen frequently in dreams. Those symbols are understandably more personal and less common and based on personal experiences in life. *Symbols,* as a common understandable signal content between individuals, *have had a history of evolutions*, which is not the concern of understanding the dreams but is used for explanation of the waking mind.

Used, Experienced, and Practiced Patterns

Abstract concepts use tangible patterns to communicate. Self-expression as the basis of communication with others, is an automatic and simultaneous urge to ensure self-protection and self-advancement. The brain is an extroverted system as well as an introverted system. To ensure the basic needs of self-protection and self-advancement, the brain needs to simultaneously countercheck the environment, including the society. Both are needed to satisfy his needs and preserve humans as they are.

The use of symbols, metaphors, and similes in self-expression or communication with others has been created by the brain's product of special signals and facilitates the conveying of the meanings by giving examples and a general (same family) combination of symbols in a pattern. The use of metaphors in communication facilitates the pattern of the meaning to be emphasized with a similar physical pattern of body shaking. When awake, the facts are expressed by sensory patterns common between people. In sleep, the activated images of memories are used for self-expression (talking self to self) using personal symbols, metaphors, and similes, which convey specific feelings.

In general, the imposing strain patterns in waking and releasing overstrain patterns in sleep both use known (saved memories) patterns to express, which are biases, the strain caused by inputs during the day.

Since a high percentage of the neural connectivity configuration is similar between a majority of individuals, strained patterns could be similar if they have similar daily needs and experiences. Even in a rare possible case, the way they interpret and express the daily experiences are different depending on their memories. On the other side, they may interpret the same but describe differently. According to the kinematic model, all interpretations and expressions of them have a strain-releasing basis. The continuous normal straining and releasing patterns and periodical synapse reconfiguration and releasing in overstrained locations are the bases of all brain interactions. If the pattern of these interactions are the same and the descriptive patterns would be the same, then people understand one another. However, because of different experiences in life, the big portion of neural growth has been done differently and is very person dependent. In this way, daily communication is far more understandable because of input and output exchanges, but sleep expressions are very personal.

Memory Consolidation

The relation between characters and events in dreams proves that although the different components of a memory are saved in different neural networks, but the general pattern would also be saved in the same location by the emotion encoding. The place for emotion encoding in a neural network is the *amygdala*.

The components of the different memories are similarly saved according to a similarity in pattern in a family neural network as declarative and nondeclarative memories in the hippocampus. A family network is a network with its higher-resolution networks covered with lower-resolution networks. It is assumed that because the neurons grow in a location in a fractal order, their network configurations also have similar patterns. Therefore, networks in different levels have a good similarity in configuration. Consequently, it is a reasonable assumption that similar memory components are saved in the same location.

The drive for the appurtenance of a scenario is the urge to release the confined energy of the strained point and, in other words, to release an emotion. However, those emotions had been made under different circumstances with a variety of character and behavior combinations.

Still, an emotion is preserved by utilizing different components from different circumstances in a scenario in random if those circumstances had produced the same emotion. It means, an overstrained location can submit releasing energy through spatial neural networks of different resolutions more randomly.

In waking time, such a random choice is not possible because there is a closed loop of processing with the input and output's strong feedback.

In short, in sleep, the brain activities are performed in a closed system and in an open loop, while in waking time, the brain activities are performed in an open system and in a closed loop. This is the basic formula that will be used to predict how the emotion will pop up in dreams and in a possible related scenario.

Pathways are passive during sleep and active during waking.

Active pathways are fed by sensory simulation or self-intentional-guiding inputs, while molecular activators like genes and DNA patterns and location metabolism accelerate the pattern of the synchronized firings, which will happen in synapses but flow over a network structure.

Passive pathways are routes of releasing stress flows initiated from tension over lower-resolution neural networks intending to relax tissues.

Conclusion: By many dream analyses based on above descriptions, *we can map a structure of memories for any individual* by extracting memory components from previous dreams with their specific core emotions and furthermore predict the next dream's possible characters and behaviors by a random combination. The only information required is what stressful event has happened before sleep and what the related emotion was.

Mapping Memory Locations in the Brain

The following is a claim but suggested to be used for programming purposes at this time:

(1) Writing daily memories in a diary (as the input), (2) writing past memories when filling a dream-analysis sheet (to complete individual database), and (3) taking note of dreams (to analyze) are the requirements to map memory sites over the brain.

It is believed that dreams, in general, compensate daily event stresses, utilizing the past memories to find a sufficient potential to consolidate the daily memories physically by plasticizing new networks. In this direction, emotion felt in the dream is common with the emotion experienced during the happening of stressful daily events and emotion experienced in relevant past memories.

In any dream, the narrative is a random blending of past memory components and the stressful-event working memory components. It seems the only rule followed is that a combination of different components should evoke the same emotion as experienced in the stressful event and as experienced in the utilized past memory. The creativity strength of the dream is the integration of different components of different kinds of memories to stimulate a unique emotion, the one that was felt during the course of the daily event. The created emotion is the result of the unfolding and refolding of the neural networks (family memories) in the same location (the overstrained location). This creativity makes a breakdown of the stressful event components and accommodates them in their right neural networks.

Memory and its Physical Stress-Strain Basis

To program a package simulating a virtual dream, a planning structure is based on finding highly developed brain neural networks (which is a big project, being worked in few different research centers), principles on how to assign a memory to a local network, the elastic behavior of different areas in the brain in a computation way, and preparing a correct classification of emotions.

The following are some detailed projects named to be refined and defined according to the kinematic model:

- Search for memory building, training, and educating theories
- Repetition and relaxing-time behaviors according to area stress-strain characteristics in the linear range
- Repetition and relaxing-time behaviors according to area stress-strain characteristics in the nonlinear range
- Search for "memory and time factor" experimental research works
- Search for the loss of memories due to location injuries by shocks or so on
- Matching of working, short-term, medium, and long-term memory creation as defined by psychology with a kinematic model way of description
- Developing working, short-term, medium, and long-term memories as described by the kinematics model
- Search for erasing memory (forgetting) theories and time factor
- Search for Alzheimer researches and unfolding phenomena by aging
- Search for studies on accidental remembering

- Possible application of pattern separation and completion methods to the mapping of the memory sites
- Search for fractal growth of neurons and possible types of firing fractals
- Connectivity flexibility and elasticity property, plasticity versus elasticity in different areas
- Finding a general model to combine synchrony in firing and stress-strain phenomena
- Clinical and experimental experiments and suitable brain-imaging techniques with short intervals

Specific Family Memory Zone

Same family patterns, which can be used as a symbol of one another, have been developed physically in one location and will be developed with similar configurations. Stimulation of any pattern in the location can be followed with stimulation of others in a pathway sequentially. Increase in new patterns, because of exposure to a new stressful event, will continue in the location with similar patterns of networks that have been configured before. Therefore, during regeneration, if a new similar network in configuration is being processed, the structure can be excited and energy flow is very localized; related interactions are batch type and with a short time span rather than the flow being fluent or taking time as waking time. The attractor is the emotion involved, which directs the random streamlines of the flow in the same direction.

Copy-Making Process of Energy Wakes in the Middle Brain

Each wake of free energy is alive up to the time the input is active or on maximum limit up to the time that its working memory is efficient. The wake converts into output for action, and expression and may be copied by excess energy in the middle brain for memory saving in a regeneration period. The copying occurs by a resonating process; it means the wake energy, which has an almost steady state, fits for the similar valleys in the hippocampus energy plan and convolutes the plan for some detail connections as it was presented in the energy wake. The convolution would be consolidated during sleep.

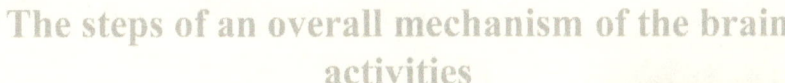

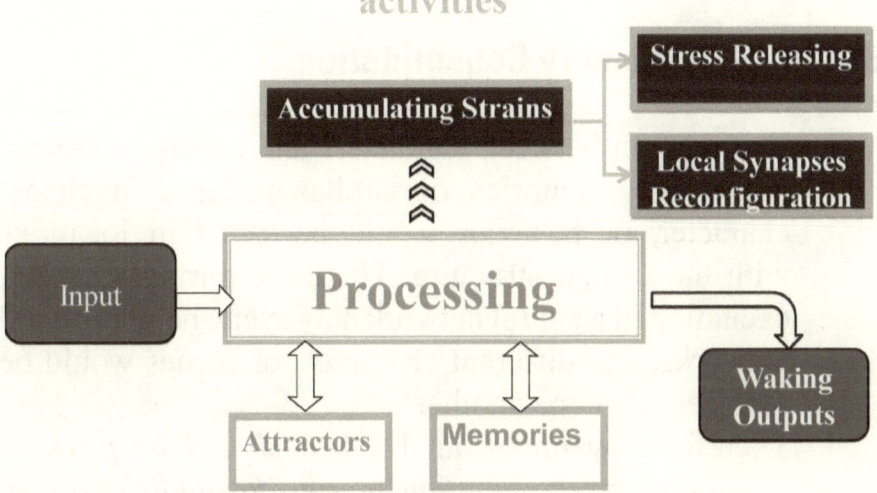

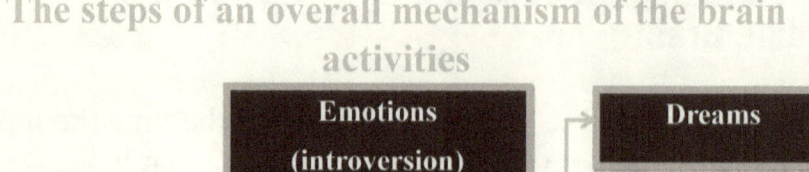
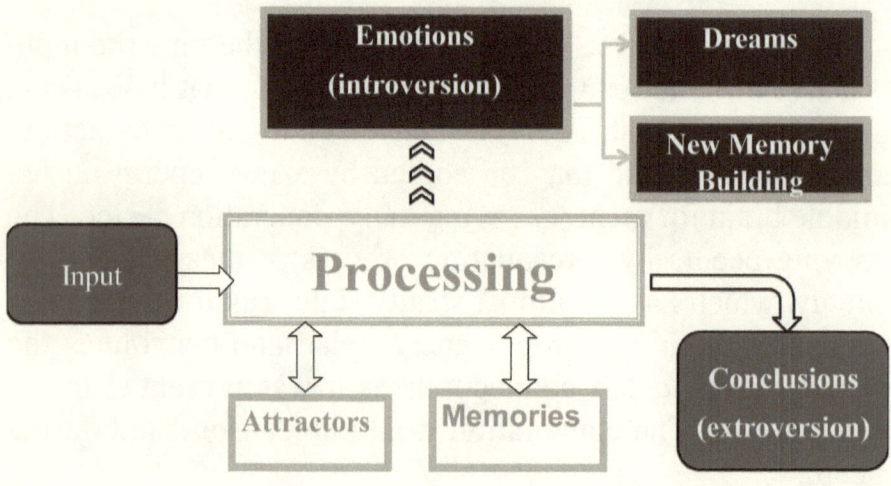

Structure of Memory Consolidation

- The relation between characters and events in dreams prove that memories of similar patterns in shape, character, or behavior would be saved in locations with the similar structure. Therefore during a dream, excitation of a neural network may excite neighborhood networks, and different characters or scenes would be replaced with one another.
- Therefore, by dream analysis, *it should be possible to map a structure of memories for_individuals* in a virtual space in the brain. The pathway of a brain activity during the day and possible overstraining of a local neural network can be determined by finding a similarity between past memories' neural network sites in a network model of the hippocampus and the

overstrained neural network site in a network model of cortex.
- Brain locations are acting passively during sleep so that overstraining urges the location to fire synchronal and are active during waking so that it reacts with incoming energy flows.

Mapping Memory Locations Over the Brain

Writing daily memories (as the input to a dream monitoring package), writing past memories (to complete individual database in the package), and writing dreams (to find a way for saved memory configuration) are the requirements to exercise mapping memory sites over the brain.

It is believed that dreams screen daily events with the past memories to consolidate the daily memories physically in conjunction to relevant memories in the past. Therefore, an emotion felt in the dream is common with the emotion experienced during the happening of the daily event and emotion experienced in relevant past memories.

In any dream, there are memory fragments of the past and images of recent events. Memory fragments, in connection with one another, convey the same emotion as experienced practically during the happening of the whole event experienced made of elements of substances, characters, their behaviors, and the circumstances they reflect. In the same way, static components of memories connected by action components of memories remind a circumstantial memory, which conveys the related feeling to the emotion. The creativity in the dream integrates different components of different memory types to simulate an emotion, which was felt during the course of the daily event. The created emotion is the result of unfolding and refolding of the family neural networks that host those memories, and because they are similar in configuration in any location and the stress flow is a spread type of flow from the overstrained point, the components are combined and provide a scenario of random but overall contingent narrative.

Specific Family Memory Zone

The same family network pattern, which can be used as a symbol of each other, have been developed physically in a location and will be developed similarly in the future too.

Stimulation of any network pattern in the location can be followed with stimulation of other networks in a pathway sequentially. The memory sites in the middle brain resonate with overstrained networks in the outer layer, accelerating the production of the dream scenario.

Therefore, during dreaming, the pathway is more localized, batch type, and with a short time span rather than fluent and time taking as brain activities in waking time. The stress releasing streamlines are from the highest strained point, rolling downward in a random route consisting of higher magnified neural networks, which are ready to react too.

Same Functionality Site for Different Character-Scenario Cases

If a psychological statement (like "Somebody did bad to me") has a physical coding of definite adapted neurons circuit in networks, and if those networks would be excited, the statement will be self-expressed as a feeling. The networks have a definite location in the brain. In automatic thoughts and dreams, the feeling would be retrieved with different inputs but the same evoking emotion. In waking, the excitation occurs with the accidental passing of a stress flow through a related network, and in sleep, the excitation happens if the network is located in an overstrained location. Often in waking time, two or more inputs are involved and excitation happens and is reinforced by sensory inputs as well as active networks ready to discharge. Therefore, the resulting emotion is not a simple emotion but a combination or a serial change in emotions during a brain activity. It can change from a negative emotion (depression type) to a positive type (a satisfaction) if releasing continues.

In short, a releasing process is mostly not a simple release but may cover networks that have an intermediate change of emotions.

Interpreting of Detail Patterns with General Patterns in Dreams

The main pathways are located on the outer layer and connected to the input-output zone, where input sensory is entered from and motor outputs go to the environment. The resonating networks that direct the pathway are attractors by nerve connections and resonating. Attractors are in different scales from genes in a molecular scale to neural network scales of different resolutions, including instinctual needs and long and short memories.

The interaction of resonating networks can be diverging to branch the pathway (analytical), converging to combine the branches (synthetic), or make two different subpathways go in parallel.

The output of the pathway could be a simple pictorial fact as it is in the environment or abstract symbols, which convey deeper layers of facts. The language zone is the developed area of the input-output zone, which is a gate for more abstract symbols. In separate areas of the language zone (Broca's area), network types and higher numbers of resolutions should help to maintain more complex patterns of firing fractals. However, the same processing mechanism in the outer layer will apply to these units in either complicated symbols or simple pictorial shapes. The relation between the complexity of input and complexity of the output is dependent on the complexity of processing. The creation and buildup of memories of abstract symbols are expected to be located in the frontal lobe.

Babies start with names and then learn to use verbs and later make sentences. Symbols, metaphors, and similes come later to group and generalize the individual meaning in statements. Such new abstracting should appear with the development of new complex consolidations in more complex structures and has had many steps by several sleep periods to learn that.

During the process of abstraction, integration of individual case memories has been generalized in location specialties. Therefore, functionality in a different location has a different complexity. Physically, neural network complexity should increase from the lower brain to the outer layer, and the highest degree of its complexity should be found in the frontal lobe. For example, if the memory of the names and events are saved in the middle brain networks, the outer brain makes stories, and the frontal lobe makes reasoning.

BRAIN FUNCTIONING AND REGENERATION

Making metaphoric concepts from simple symbols in dreams shows how a description of highly complex patterns can be done by simple patterns.

Overstrained locations normally happen over the outer layer of the brain.

Dream Automade Narrative

A dream does not have any message by itself. However, it acknowledges a memory that is being built from the daily stressful event/s in the same location where emotionally relevant past memories are saved. Stressful events are categorized as those that have a strained impact beyond their linear relationship with input stresses.

Any symbol in a dream, like a character and behavior, has an emotional attachment and a following feeling, which would be expressed with related images. The overstrained location excites a similar structure in the amygdala, and later, structures resonate randomly with declarative and nondeclarative memory components that combinations of them evoke the same emotion.

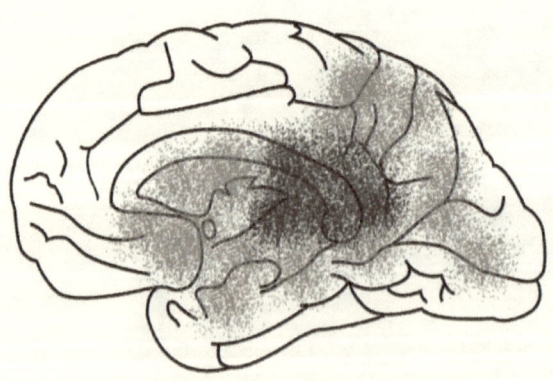

A collection of symbols and emotions will grow by life, connecting common feelings and images that build the language area and developing by the meaning of metaphors and similes. A dream narrative and scenario is randomly selected from the collection, and the spread of the releasing stress flow is very chaotic along the collection.

Dream Fictions

Dreams are full of metaphors, similes, and personifications around attractors of emotional theme. It starts with a necessity to release from the stress of a location with the highest strained situation and moves to other strained points for the next episodes. The strained points can initiate releasing individually or in combination. According to the strength of strains in different locations and pathways, the REM and NREM narratives and memory consolidations of the night would be proceed.

There is a climax of releasing in the beginning or later, which can interrupt sleep as a nightmare or can continue to further release stress, which gives a pleasant, satisfactory, and joyful feeling.

Unfulfilled or attacked needs during the day are painful and stressful. They strain some locations severely. Regeneration is required to remove those strains, and the removal is accompanied with a virtual fulfillment according to the neural networks excited as memory sites. Interruptions, because of high climax points or external noises in dreams, make the sleep inefficient in recovery and regeneration.

Highly strained locations initiate a reconfiguration performance as new memory building in a NREM stage. The stressful events will be consolidated during this stage by sudden change of configuration to balance big differences around the highly strained location. NREM will be followed with a REM stage to equalize and make uniform stress distributions fully.

Because the starting of an NREM stage is related to the highest strained location, the following NREM stages are reduced in strength. In the same way, REM stages are accompanied by a softer and softer emotion expression along the sleep duration.

Areas of the brain are divided as per circuits and functionalities made by time. These functionalities are formed according to local strain-stress properties.

During a free energy flow in a pathway over the outer layer, lower contrasted level of the neural networks resonate with the memory networks in hippocampus; while middle contrasted level of the neural networks resonate with networks in the amygdala, and higher contrasted level of the neural networks resonate with networks on the frontal lobe. In total, all kinds of components that make brain activity meaningful would be available to create a dream.

During waking, initiations start from the synapse level in the input zone and terminate with command signals to motor outputs.

During sleep, the process is reversed and initiation starts from strained networks in the layer, transfers through different resolution levels of the layer to equalize the energy, and terminates in the synapse level configuring firing fractals.

By this process, confined strains on the layer unfold in dream narratives made of equivalent images to firing fractals, which are chosen randomly from components of family memories, which convey the same emotion.

Symbolic Expression in the Dreams

Patterns of the synchronized firings are the creators of the characters, scenarios, and all other elements of the narrative for a dream.

Patterns in shapes of firing fractals enter, travel, and exit a pathway, when strains are imposed during travelling on existing synapses configurations and make synapses to deform. Therefore, all types of minding appear as patterns, and patterns are the only media the brain works with.

Patterns develop as images, dreams, feelings, decisions, interpretations, claims, beliefs, words, figures, and statements. Patterns develop emotions as love, hate, and so many others.

Patterns are expressed in different ways of expression like paintings, poems, assumptions, games, stories; and make history by plans of wars, freindships, innovations, designs, and so on.

Even in laws and ethics, some concepts and instructions are gathered together to make a pattern.

Patterns move as ghosts and clouds as well as streamlines in materialistic bed streams; and finally, patterns physically appear as structures, configurations, a regular flow, an internally chaotic flow but regular in general, and a turbulent flow. In the latter, eddies, drags and wakes create extraordinary shapes. Patterns do not differentiate between wave and matter. They need only a few basic parameters to be defined like frequency, amplitude, and spin.

There is a static pattern of neural networks and connectivity of those as patterns inside patterns in the brain; however, input patterns enter in, settle in, and go out of brain when the outside of the brain creates patterns of expressions, actions, arts, and real productions.

In short, substituting material media with symbols and lines, the essence of their continuity would appear. The essence of continuity in material in the environment would be replaced with the essence of continuity in stress flows in the brain and will be represented by synchronized firing energy streams as the mind. The active patterns in the brain are made of firing fractals and the flow of eddies, energy pockets, and wakes.

Reality and Thoughts

Real stressful events come in as ghosts, challenging sleeping memories with definite elasticity rules of brain layers and exiting as imaginations or actions.

If physical brain rules would be screened by experimental tests, the product patterns would be deterministic as scientific facts; otherwise, the product patterns would be interpretations, views, claims, assumptions, and dreams.

Patterns in the environment are very restricted to the material and energy they carry. Long-term evolution is the only way of relieving from the previous limitations and emerging in a new developed pattern, but in the brain, they can be shared between different kinds of mind existences. This fact is the strength of the brain—to be able to create many kinds of fractals by synchronized firings and make many combinations of them in the brain when it cannot happen in nature with many possibilities as in the brain.

Another essential uniqueness that the brain has is its flexible elastic material, which can shape itself with a variety of changes and again find its original identity. When there is no input and output, previous patterns disappear and firings dance in Brownian fashion and keep some other parts of the brain, which is not active, at rest.

Reality is melted in the brain, crystallizing in a pseudoreality. After being processed, it can get out of the brain as a new reality or a personal view.

Imagine how a pattern can change before coming into the brain in reality, how it is processed in the brain, and how it is when it goes out of the brain:

- A nude girl jumps into the water, spreads as small fishes dancing in groups, and unifies as a pariah.
- A tree breaks in pieces of small birds flying in groups and unifies as the moon in the sky.
- A nude man goes inside a whirlpool and dissolves as streamlines being deformed by jet waters on the walls and comes out as clouds of steam from the steam room, creeping from the door opening.
- A competitive colleague attracting the boss is dreamed of as a teaboy.

Dreaming Principles

A. The key to trace dreams against daily events is shared emotion and consequences, feeling in the time of *event dream content past memory*.

However, the intensity of the emotion is very deep in sleep because of the absence of inhibitory thought, which is present in waking.

The common emotion in the event, dream, and a related past memory is due to the order and locations that memories are built. Memories with the similar patterns would be saved in locations with the same structures. Normally, the type of configuration in majority of neurons and neural networks is the same in one area. It is expected that the memories or memory components that convey the same emotion are built next to one another. When a free energy transfer happens over a pathway in the outer layer and an energy substrate on that pathway during interaction finds a temporary pattern as a memory site network, they resonate with each other and take part in that activity. Therefore, they find the same emotion in the moment. This way, the event as a working memory and the past memory together evoke an image with a common emotion.

> B. By resonation process, the brain is so flexible in preparing a situation that the stressful sources choose a scene or a scenario made of the same family of symbolic expressions, which convey the same emotion.
>
> - Strained locations can evoke positive emotions, appearing with the inspiring of satisfactory attractors, or negative emotions, appearing with the absence of satisfactory attractors.

The Logic Behind Dreaming

According to the model, as it has been developed up to now, if the neural network configuration in the brain, especially in the hippocampus and amygdala, would be clearly known and the neural networks on different areas of the outer layer including the frontal lobe is known for the structure and configuration changes due to layer elasticity, the resonating process between those locations can be analyzed. Assumed samples are given as follows:

1. Outer layer area neural networks
2. Overall brain neural networks

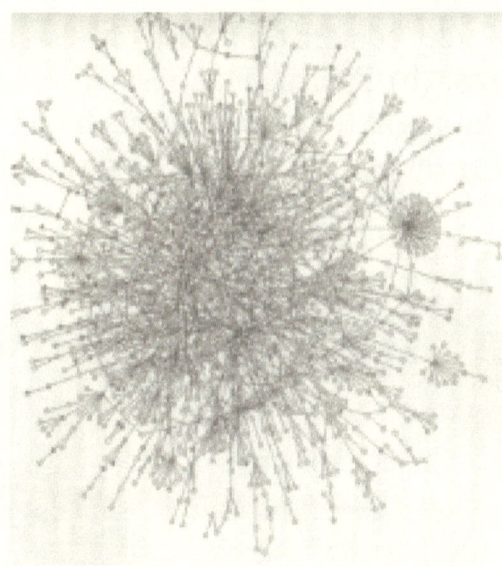

An illustration of the input-output zone, which covers crowded branches of the same family memories for programming purposes. (Recreated from *Branches* by Philip Ball)

3. Areas cover trees of the same family patterns of sensory inputs

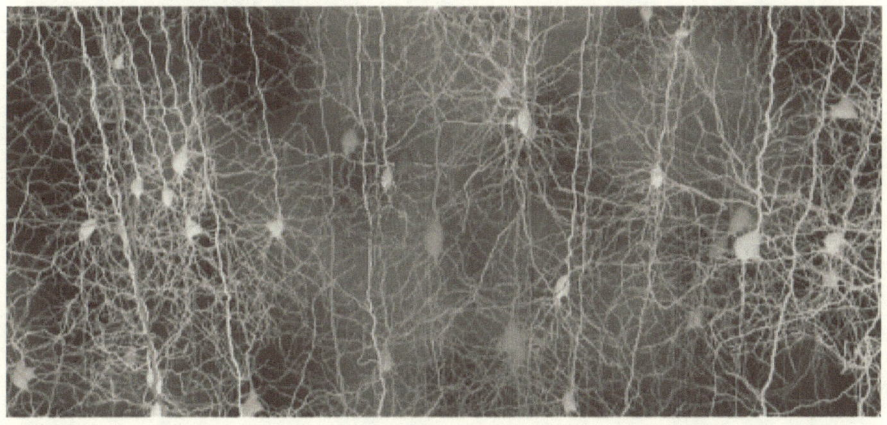

(Reproduced by permission from *PLoS*, "Blueprint for the Brain")

4. Each tree contains process levels of (a) overall patterns, (b) intermediate-resolution-level patterns, (c) high-resolution levels of thin branching patterns, and (d) the synapse level as site of firings. Each collection of synapses presents an image of a sensory input consisting of things and events.

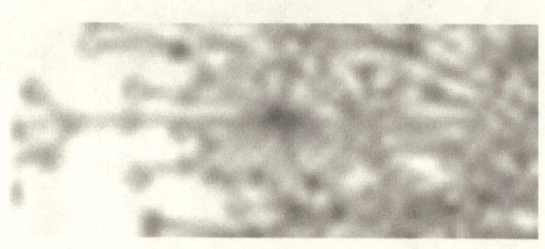

Patterns with similarity in configurations demonstrate a different but similar pattern things.

A family of close-meaning symbols and metaphors

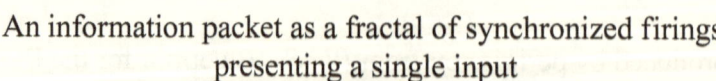

An information packet as a fractal of synchronized firings presenting a single input

5. Each sensory input can resonate with most similar patterns. If it is exactly the same, it is perceived as a known substance or event. If different, its relevant stress flow will take a working memory to be saved if it will overstrain.
6. Each tree, when excited in part, creates the same family images or symbols in high-resolution branches, the same family statements or metaphors in intermediate-resolution levels, and the same functional patterns or similes overall.

    6.1. Elasticity behavior and the remaining confined strains in the location determine the time factor of creations.
    6.2. Elasticity behavior and the accumulated confined energy in the layer for longer periods of time determine the sleep type of dreams in general.

7. In waking, sensory inputs initiate stress flows over the layer from high-resolution levels to lower-resolution levels. Therefore, the images and perceptions are ahead in any stress flow presented as feelings or currents of thoughts.
8. In sleep, stress relieving starts from the layer going to the high-resolution and synapse level to create images. The stress flow follows the layer elasticity behavior. Therefore, the emotion drive creates the images and is in head of images. Consequently, the images are not determined and can be any image from the same family that evokes the same emotion. These images are symbols of each other.
9. Elasticity behavior (strain versus stress) formula is to be adopted for each layer and area.

10. Accumulated confine strains remaining are accumulated during waking time and therefore can be shared between different brain activities during the day.
11. Release of the accumulated strains is released during sleep and may blend different stressful events and not segregate them.
12. Releasing in starts from a low-resolution level in the layer and flows toward creating images.
13. According to the above direction of the lease of free energy, an integrated image as a scene or scenario is expected to be highlighted ahead of the characters.
14. If the confined strains are not strong, positions on the location return back to the origin immediately.
15. If the confined strains are strong, the elasticity behavior would be nonlinear, and a residue of the confined strain can remain trapped in a location. Having frequent pathways activated during a day over the same location of the brain can either reinforce or nullify the strength of the confined strains in any location.
16. The location or locations with picked confined strains may cover different family memories.
17. During regeneration period, the strains would be distributed in two directions across the zone and quickly release as network deformation or uncompleted release according to the elasticity property or required activation energies on the location. If any network deformation would happen, some memories may be erased and new ones created.

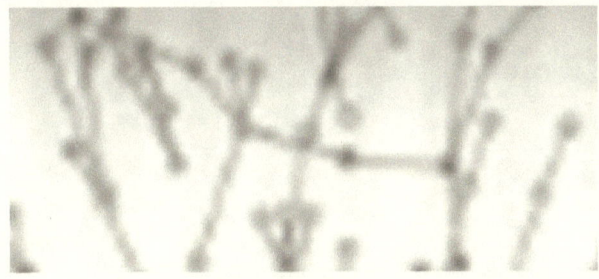

More than one family memory in an overstrained location may create a blended dream.

18. Naturally, the stress release should start with the most strained point and continue further to less strained points. However, such a sequence could be chaotic and not in order.
19. Each picked location of the strained points conveys the related emotion. If two overstrained locations start to release at the same time, it can be a lucid dream.
20. The regeneration process is a combination of reconfiguration and distribution processes. New memory building would be located inside a family memory network but can be shared in two different family memories if corresponded to a blended dream.
21. Images can be in three types of stable firing fractals as characters—generally stable but irregular internally as behaviors and confused if irregular.
22. In a dream, characters may be known by names, feel as one of the members of a group like family or colleagues, or seem generally familiar, which also depends on how the firing fractal is stable in a definite level of network resolution.

A Dream Package Design

A dream is a mirror of daily feelings and how we believe in our heart and how far it is from our desire!

The induction of a stress flow as a feeling during the day may build an overstrained location if highly stressful so that the location coordination in its stress-strain relationship goes beyond the linear range. Then the releasing process of the strained location appears as a dream.

Any area in the outer layer of the brain is formed with similar neural network structures, which evoke similar *emotions* due to the locating of a newly created network in a location with similar networks. Those networks could be formed by the shock of very highly stressful events or gradually repeating events or a learning practice. However, the components of memories would be saved in the middle brain, distributed according to their required structure of networks. Therefore, in the case of a definite emotion, evoking a collection of memory components consisting of the characters, behaviors, and the scene are required to fulfill that emotion.

Dream analysis helps in studying the brain activities during regeneration periods in a large scale and can cross-check the location of memory neural networks where they are saved and related type of firing fractals (if it is known) as well as the study of free energy transfer through the layers in the future.

Firing fractals and free energy transfer routes in connection with synapse's configurations are the mystery, which can guide us to determine nature of the brain activity, including dreams, feelings, and thoughts.

Strains determine which gates, in an instant, are decreased in size and would not fire and which increases in size and fire. Therefore, it determines the synchronized firings, which occur in any moment.

Synchronous firings in a pathway cause strain forces on the route, and strain forces on a route indicate the sequence of firing fractals.

The relation of a firing fractal and energy transfer in a pathway is having synchronized firings in a location and sequential locations firing. A high intensity of energy distribution determines which route can be considered as a pathway.

Populated synchronized firings induce strains. An accumulation of strains acts on the elastic material of the brain, causing its stiffness, and next is the releasing step of the elastic layer. During regeneration, deflected elastic tissues are recovering; the route starts to return to normal by releasing the confined energy and creating clouds of new synchronized firings, exciting the related neural networks. Outside stimuli cause inside stimuli during the day and inside stimuli cause the emergence of images that had been separately saved before, according to the brain's special elasticity.

The brain has a curl configuration in the synapses' arrangements everywhere. The maximum strains will be absorbed in the input-output zones. The way it releases the stresses and the routes it follows to distress are determined by the sequential changes in fractal growth in course of a brain activity. Therefore, the complexity of pathways and the complexity of fractals and the complexity of output discharging energy are related.

The brain's shape is made of the quick growth of lower-resolution neural networks during developing ages and gradual growths of higher-resolution networks afterward. Growth in new neural networks changes the elasticity slightly, and changes in elasticity determine the change of pathways, which most probably could happen.

The Emotion Generator or Code Producer

1. The input section of the DreamPack uses a language processor to recognize the emotional words in the diary across a comprehensive list of emotions when any emotion in the list had been assigned a code.

    The code given to each emotion depends on the categorization of emotions and the coding system. After an initial searching of the emotion code along the emotion categorization tree, the search will be further continued for emotion gradient along the daily note according to the emotional load of the words.
2. The past diary memories have been analyzed by the language processor for the characters, behaviors, and scenes involved, as well as emotion changes experienced during that memory. The sets of characters and behaviors and relevant scenes are listed against the emotions. In this list, a weight factor would be assigned to any of these components to show how much any of these components carry the emotion load.
3. The list of characters, behaviors, and sense sets will be multiplied for replaced characters, behaviors, and scenes so that the combination of these components, with consideration of weight factors in the recent expanded list, provides the same emotion code.

4. A data bank of the character's shape and indicative behavior/s and the type of places where the memory had happened would be prepared. In preparation of shapes, indicative behaviors can be produced in avatars with emotional symbols attached.
5. A set of characters, behaviors, and scenes from the expanded list in the data bank will be prepared in a slide.
6. The emotion code will select randomly a slide with the same emotion code.
7. The slide will be developed to a video clip according to the emotion gradient and the strain releasing character of the location.

In short, stages to produce the product are as follows:

1. Language analyzing processor
2. Entry location in emotion-tree processor
3. Defining of the strain-releasing behavior of the location to rest condition and resulting scene-characters-actions code
4. Retrieving of the saved video clips
5. Video clip preparation

Detailing the stage 3 (code producer):

For the simulator to recall scenes, characters, and their actions in a dream monitoring program, a code is required where its elements pick up the relevant saved images (in a very detailed work) or clips (in primary works).

The code theoretically is made based on two principles:

1. Shear stress (τ) changes from maximum to near zero on the location.

2. Strain versus time (ϵ versus t) would be formulated at least for different regions of the outer brain.

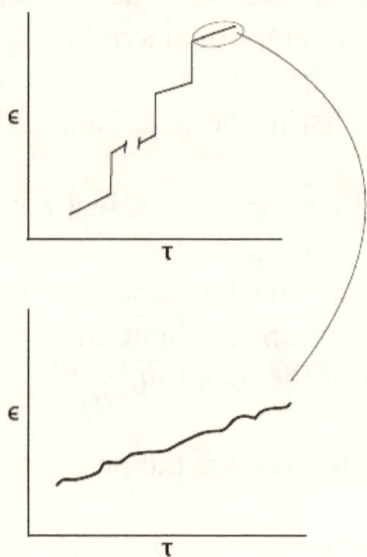

The decrease of the strain versus time is expected to be in steps as, theoretically, it is known for the viscoelastic materials like the brain. However, each step would be according to an overall sloped curve but consisting of hills and valleys of intermediate unfolding or releasing process; and each hill or valley provides a configuration of the related firing fractal equivalent to a character or behavior component of a memory pattern.

The presentation according to unfolding function (which is defined by stress-strain correlation in the location) recreates the dream consisting of images, motions, and the scene.

Mechanism of Dreaming

Memories for everybody as personal experiences contain isolated or serial events. Mostly, the isolated type is built by a stress shock and serial type by a repeating experiences.

Symbols are mainly personal depending on characters, behaviors, and the scene in the experiment; therefore, a symbol has an attached emotional load, positive or negative.

A personal symbol can become a common symbol if it evokes a common image in different people's brains. Depending on the degree of popularity, symbols become a media for communication.

Many personal symbols are based on physical configurations the synapses can find in the brain, while common symbols are limited to having shared emotional or learned meaning (same effective neural network configurations).

The mechanisms of categorization are the following:

1. It is assumed that working memory physically is associated to a wake of the confined strain pattern which still is not consolidated as a permanent memory,
2. A working memory keeps the hosting networks excited.
3. Strong working memories will end in a memory consolidation in a homogenous network structure with the similar memories.

Emotion is physically defined as an overstrained pattern. The number of emotions is equivalent to the number of neural network structural types in the amygdala. Therefore, daily stress flows are indentified by the number of neural network structural types in the amygdala. The relevant release also evokes the same number of emotions. During a release, each energy packet resonates a similar network configuration in the language area or, in other words, relieves a relevant symbol. To balance the layer from strains in a location, the energy associated with the strains would flow and evokes a symbol in the language area. The symbols are dominantly in visual form and less audio, but will form according to the sensory region involved in the pathway.

The general pattern for different dreams can be the same when the emotions felt in relevant daily events are the same although they borrow different sets of memory components from different times.

When the experienced emotion is the same, a daily working memory finds a way through past memories in a dream. The common emotion is a link between memory components and the daily event components so that the daily working memory retrieves the past memory components, either in waking or sleep. Except in waking, they should be identical as much as possible because they would be screened with input sensory, but in sleep, they should be emotionally identical because there is no input sensory to screen them. In the first case, the event components influence the pathway as much as emotion does. In the second case, it is only the emotion that shapes the pathway.

Dream's Rules

The characters or scenes in a dream may change from one to the other in the memory if the evoking emotions are the same. These kinds of characters are a symbol of one another and are sited in neural networks with the same configurations. Majority of similar types of networks are located in a same area. Therefore, the same family characters should have been stored in the same area or resonate easily with far distant networks.

The three main rules that create a dream are the following:

Rule 1: In a dream, scenes are lengthy and changes between scenes happen less than a few times, like the number of stages in a theater when changes between scenarios are fast and sharp. But unlike in theater, a dream is less dialectic and more visual.

Rule 2: A dream is composed of symbols, metaphors, and similes completes a meaningful self-expression.

Hardly two dreams are alike in scenarios and characters, but they frequently have the same core of emotion. Dreams are created and directed on time according to the chaotic energy releasing routes from a strained location and are not pre-planned.

Any dream has a product of delight, acknowledgement, a satisfaction message, or advice for an action (progressive or preventive). : A dream is a language of self-expression. Awareness in it is because of the emotion conveyed by it.

> Rule 3: Although the scenarios for a daily stressful event, a past memory, and the dream itself are different, the scenario in the dream borrows its characters and other components from the daily event as well as past memories if the their combination conveys the same emotion. A summary of the above rules are as follows:

- Any unit in the environment is made of a geometrical pattern attached to physical properties, with each property being a dimension of that unit and imposing a feeling ended in a combination of all those feelings in an emotion.
- Any unit in the brain is made of a synchronized firing pattern attached to symbols, metaphors, and similes, each being a dimension of that unit and simulating a feeling coming out of an emotion source when a symbol represents the physical property in an abstract way.
- Any geometrical pattern during the process of sensation finds an equivalent of the synchronized firing pattern (as input to the brain) and an equivalent consolidated structure (as memory of the sensation) that when it would be excited, it retrieves a substantial, eventual, or functional synchronized pattern as an image.

A Dream as a Play

Each dream narrative has two main elements of character and action (play) in it.

1. Character/s:

- *Almost all dreams are directly or indirectly self-oriented.* The indirect self is a nonself character who metaphorically represents the self—for example, a family member or a person of the same nationality that the *self* has a close sympathy to or shows a kind of emotion toward. *In physical terms, such a character has a strong resonating natural frequency with the overstrained structure.*
- *Emotion felt in a dream is the same emotion that has been felt during a recent stressful event, and the same emotion had been experienced in one or more related memories in the past.*
- *The natural frequency of an image* is one of the indications of emotion in a dream associated with a character or a symbol. It is a combined frequency bundle of the neural network, which supports the saved memory when excited.
- *Several images can have the same natural frequency.* Therefore, any of them can be randomly picked up or

substituted during a dream while it does not happen in waking time.
- *The resonating of a memory of the same emotion experience with an ongoing image happens when their natural frequencies are equal or multiple in numbers.* A related saved memory with a higher number supports the ongoing image stronger. This fact reduces the numbers of the images, which can pop up in a dream. The retrieved image would be translated to a symbol in the language area if it would appear other than an image.
- In general, the *numbers of the main category of emotions, according to various classifications, are limited and mainly not more than fifty*, while images with same attached emotion can be many. Therefore, a symbol may represent many similar patterns in the brain.

2. The play:

- A play in a dream follows a narrative, which tends to nullify or remove the strains as much as possible an overstrained location in a layer by satisfying interactions between character/s and unmet self-needs and wants.
- A play has a distribution of the frequency within a definite range, which links the natural frequency of the characters in combination with the frequency of the characters' structure; it provides the natural frequency of the overstrained structure.
- In an energy plan over the outer layer, characters are located in valleys, and plays are represented by a slope of hills.

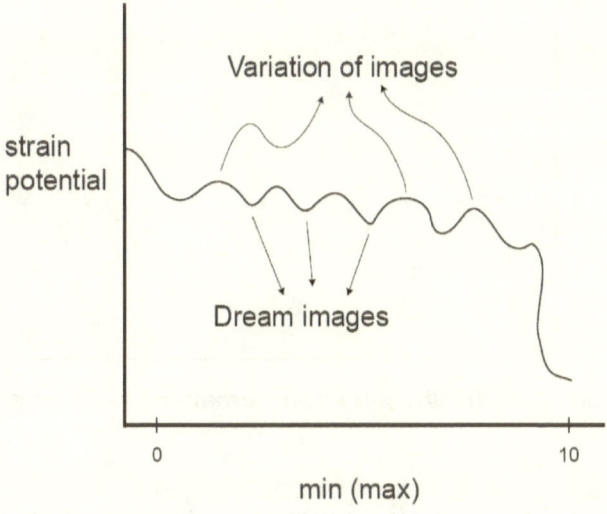

- A dream normally should end in a lower strain level in a dynamic energy plan. However, there are dreams that end with *exploding emotion and are called runaway dreams, like nightmares,* and there are dreams that the strained pathway continues to oscillate if it would not be interrupted by awaking and strain would remain in a level as it was before sleep.

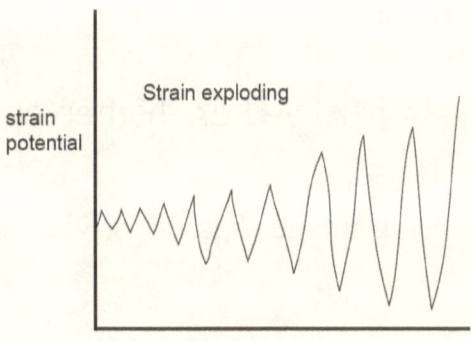

A nightmare is a "strain exploding" phenomena

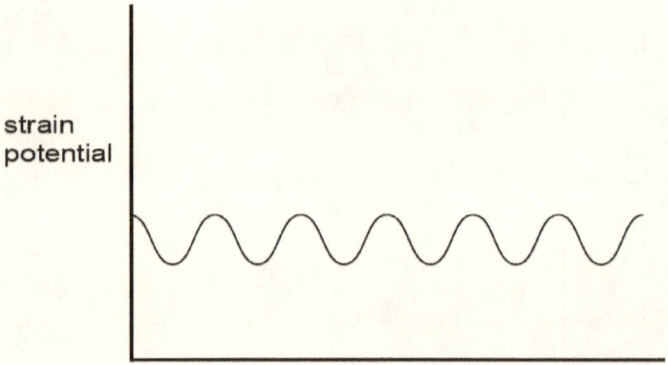

An unreleased dream is a strain remaining phenomena

- The relaxation efficiency of a dream depends on the *elevated strain-stress characteristics* of the location, which directs the play. The *elevated strain-stress characteristics* are related to the overstrained condition while the *general strain-stress characteristics* are related to the relaxed condition of the brain.

Characters' Images in Mathematic Description

A general form of energy fractals when appearing follows the following formula:

$$\partial A(\tau_x)/\partial(\tau) = \lambda A - \mu |A|^2 A + K\Delta A + \text{higher orders of } \Delta A$$

A is the appearing image, λ is the driving stress strength, and

$$K\Delta A = K\left(\partial^2/\partial\eta 2 \times \partial^2 A/\partial\tau + \partial^2 A/\partial 3^2\right)$$

is the dissipative term. If *K* is zero, the image is stable and no releasing is performed. If K is not zero, depending on K and λ magnitudes, the image would be terminated appearing a new image; or a runaway interaction will explode the image as a nightmare. In general, the driving term as a reacting function and the dissipating term as a generating function interact together. When they are in a balanced state, the order as an image is maintained, and when unbalanced, an image appears or disappears as a result of the interaction.

"The fractal can have the spiral, wave, and hexagonal patterns or topological defects and forms of turbulence" (*Motion Mountain,* vol.1, 323). The character or behavior components define the property and type of emotion attached to the image.

Turbulence is a range of chaotic motion that orders parameter increases to a high valve.

In the formula, two terms of driving and dissipation changes have a main rule. *The dissipation term is the term that determines when the stress relieving is convergent, continuous, or exploding as a nonstable term.* An image would be created when driving stress and releasing stress are in the same order. The stress-strain behavior of the brain is not linear in terms of input stress. Consequently, the location as a chaotic media is suitable to create an ordered pattern of firings in a fractal shape.

Distinctions Between Characters and Play

- A character is attached with an emotion while a motion is attached with a feeling. In other words, feeling is a dynamic state of an emotion.

- The difficult part in a dream-monitoring project is to define and formulate a pathway with the elevated strain energy and the pathway changes.
- The trait part of the basic strain energy can be analyzed by an instruction book, documenting the most stressful events in the life. Preparation of a more complete, customized symbol dictionary and how much these symbols are meaningful for an individual helps to improve the program for more precise dream estimation.
- The elasticity property of the layer can be extrapolated from the stress-strain relationship for the dead tissues, which are already available for the relaxed condition.
- However, the evaluated strain condition is a subject to be studied more in the project. Such a study helps to evaluate the probability for the estimation of which a stressful event in the day may create a dream and which will come first. Although it could not be the fact, but for easiness, it is assumed that the location with the highest strain will initiate the first dream and the rest of dreams will be associated to releasing of the other locations according to the strain strength.

Cognition of Static, Motional, Circumstantial, and Emotional Images

- Normally, the perception occurs for a set of substances and their relations together rather than for an isolated substance. In addition to a set of substances and their motions, a background scene also supports the emotion, which is attached to characters and actions.
- Memory saving of a substance or an event happens in sets of components connected with their behaviors or actions toward one another. Suppose that the set can be planned as energy hills and valleys in a plan when each valley represents a stable condition of a shape like a character; the slope in the connecting hills, which represents an unstable condition, will reflect the motions; then the valleys and hills are surrounded by many other field curvatures, which make the background scene. Such an imagination provides how a memory can be saved in different shapes of neural networks.
- The location map with all the related valleys, hills, and scenes in a relaxed condition has a difference compared to the location map when it is overstrained because confined energies change the curvature of the hills and valleys slightly.
- The same sensory inputs would have the same mapped valleys and hills, and this map should be the same as experienced similarly before saved as a memory of them. Only in future experiences would the map be perceived as a known input and strengthen the related memory.
- Different sets of substances or characters and their motions or behaviors would be saved as a space of

energy planes with many crossings. The crossings in plans (or nodes if they would be imagined as networks) are the locations where characters are saved in the layer.
- Suppose the plans are dynamic and have changes in convolution, then new creations (of characters and moves) will be imagined as not having any reality. This is the assumption that happens during a releasing process of an overstrained location.
- The middle brain, including the hippocampus and amygdala and other parts, are spaces for static memory-saving plans (declarative memories). In the outer layer, the spaces of plans are far dynamic in curvature changes because of higher elasticity.
- Any local plan in the planned space in the outer layer during changes may find the same configuration in the middle or inner brain. In that case, the local structures resonate together, refreshing an emotion as the resonating factor.

The Combined Pattern of the Releasing Networks Create the Overall Pattern of a Dream

The locally continued energy transfers are located on the outer layer from the input zone toward the zone. The reactants that direct the pathway are attractors, interacting from a far distance by nerve connections or resonating far nerve connections. Attractors are from different types of genes, procedural networks, and solid networks of the long and short memories. Interactions between transferring energies and networks in the pathway can be analytical, synthesis or parallel, depending on the pathway type consisting of diverging branches, converging branches, or parallel branches. Any chain of interactions in the course of a pathway can have a transformation in the language zone to be expressed. The language zone is in conjunction with the input-output zone. In separate areas of the language zone (Broca's area), complex patterns of symbols of individual substances, events, and functions are combined to be expressed. The resulting pattern is a symbol instead of a signal command for action.

Symbols in different shapes are treated as objects themselves when entered as inputs to the brain. The product of the processing can be a symbol with higher abstraction. The creation and buildup of symbolic memories are grown in the frontal lobe. Babies start with names of substances then learn to use verbs and state in meaningful sentences. Symbols, metaphors, and similes later will be combined in groups and generalize the subjects. Such new creations would be saved in new consolidations of memories in a more complex structure of the language zone during frequent sleep periods and will be processed and developed in the frontal lobe day by day.

The whole process of abstraction, the substances, and the motion memories convert to symbolic memories; and their processing from the outer layer develops to the frontal lobe. Therefore, functionality in the frontal lobe has more depth or complexity in abstraction than in outer layer. Physically, functionality power increases from the inner brain to the outer layer, and the highest degree of its complexity is formed in the frontal lobe. In other words, more static simple functions for input processes develops for higher processes of feelings in the middle brain; it develops for thoughts and storytelling in the outer brain and develops for reasoning by abstract tools in the frontal lobe.

A coordinated processing of the outer layer and frontal lobe in dreaming pathways sometimes cause the *interchanging of simple symbols* (names, events, or functions) *with abstract symbols in dreams* in a metaphoric way. In other words, the description of highly complex patterns is done by simple patterns. Strained locations can happen in any substrate network in the input-output zone, outer layer, or frontal lobe. However, overstraining happens most frequently in areas related to visioning. The confined energy of the strains will be consumed and consolidated inside the brain because of the closing of the output routes during sleep. The uniformity or distribution of energy through the input-output zone including the participated areas of the frontal lobe convert the pattern complexity into an abstract pattern of a straight symbols.

Similar overall network patterns in the frontal lobe and in the outer layer resonate together, using simpler functions in the outer brain in a more complex functional network in the frontal lobe advancing the creativity.

Remembering a Dream

Except in cases in which a dream is interrupted, it would hardly be remembered.

In case the dream is interrupted, it may be remembered because there still is a remaining confined energy for retrieving the rest of the dream. Early morning dreams, especially, are *easily* remembered because they are interrupted by waking. The same happens for nightmares because of emotions exploding and being awakened by them.

A Circumstance, Brain Processing, and the Emotion Involved

Having more integrated neural networks in the layer level causes the functionality strength to increase from the synapse level to overall layer. The functionality strength is also highest in the frontal lobe, less in the outer layer, and far less in middle and inner brains. Therefore, involvement of any region in the overall brain activity takes the benefit of different functionality integration.

Assuming that an input is a set of components, a circumstance is a meaningful combination of components interacting with saved memories when targeting a need. The meaning of the combination is the emotion the combination emerges with. Similar circumstances are recognized with similar combinations for the same emotion when components in the combinations are varied. Emotions are limited to types of combination types that can happen during processing over the layers, and the combination types are limited by the stress-strain property of the layer.

As configuration complexity increases from inner to frontal lobe, their behaviors also change from more instant in the inner brain and a behavior's time-dependency increases to the outer layer and frontal lobe according to the elasticity of their networks.

A real circumstance, the distribution of firing gates during reading the circumstance and a memory of the circumstance, should have a common physical explanation; and the emotion shared during reading or retrieving the circumstance is an index for their physical structures. Therefore, the virtual circumstance that a new memory would be created in a dream should appear in a scenario that refreshes the same emotion. The narrative for the scenario selects randomly the characters from previous scenarios with the same evoking emotion.

A memory is about an unwanted, unmet, or out-of-expectation circumstance in respect to the self's needs and wants. The related location for the memory had experienced a strain and therefore an emotion before and during the consolidation. It means the emotion is a representative of the circumstance while the related circumstances for the same emotion can be several. If, during waking, an actual circumstance evokes an emotion, then during sleep, an excited emotion creates a virtual circumstance from saved memories. The core emotion of a brain activity indicates a temporary configured network in the outer layer because of a stress-flow interaction but indicates permanent networks of some previous experiences in memories too. In this way, it can connect those networks by resonating. A new network can be build during sleep with the same configuration and, consequently, the same emotion.

Therefore, the experience of memory network building also evokes the same emotion. The temporary and permanent before-and-after regenerations are not so different; they are just different for the reconfiguration of local synapses as a fine new memory network to establish energy balance again. Therefore, the past, forming, and recent days' memories with the same experience of the emotion should have been configured similarly in their networks' pattern.

An experienced stressful situation that had imposed the stress in the brain may not initiate a related dream in the same night, but in following nights. It is probably because the sleeping time is not only due to stress saturation but because of daily periods as well. Therefore, the distribution of releasing is not in a simple order.

A *circumstance* is the space containing components that, in combination, create an emotion but individually have also taken part in a scenario providing that emotion.

Although the self is the target in all behaviors, the self can be extrapolated to self-originated things or memories like belongings, relationships with close connections with self.

Regeneration is rough daily and not exact; stressful events of a few days before may become a subject of dreams.

The Song of an Emotion

Dream is a personification of the emotions, having an emotion has experienced similarly a in a past event as during a recent event.

A dream gives personal characteristics to emotions by personal experiences, which are rarely common between people. The characters and scenes are symbols, which appear in the dream to convey the same emotion that reflects from a strained location of a voice suffering. The strings of the structure play a familiar song from the past that was heard recently too.

All regions in the brain are graves of sleeping feelings, which are reflecting the killed desires or unexpected fulfilled desires. They have been buried in different locations according to the structures they have. If a new similar strain would be imposed on the location, then the excess of strains gives life to the dead feelings to simulate a play of characters in a special scene. They refresh the same smell that was experienced years back by shock or training or special experience.

Emotional, Static, and Dynamic Elements in a Mentation

The amygdala in the limbic system is a code panel for the emotions where relay switches (codes) for memories are located. The emotion relays group the static and dynamic memories in separate neural network configurations. The type of structural arrangement in it is such that any same-family-emotion codes are determined in a mother network and have a hierarchical order of networks in it. Any network structure is to tune specific declarative or nondeclarative memory sites in different areas with an ongoing neural network configuration in the outer layer. It is assumed that a similar panel code for the tuning of structures exists between the inner and middle brain as well as between the cortex and the frontal lobe, which will be discussed in another section. Codes in each layer are considered as attractors in the other layer.

Elasticity limits in any layer is the boundary for hosting the category of images that the layer can emerge with or save.

In short, each image creation in the brain is attached to an emotion. When the amygdala is damaged, emotions are gone and past memories cannot be remembered.

MOSTAFA M. DINI

Recommendation for the Basic Programming of the Dream Monitoring Package

The family configurations in any area of the higher brain (cortex) are saved as a code structure in the lower brain (middle brain). The connection between these two is made by long vertical neuron connections between areas and by the resonating of structures when they find the same natural frequencies.

Resonating happens between a memory structure and its similar ongoing structure in other layers when the structure code as an emotion carried by a specific structure in the amygdala confirms the similarity.

The whole brain can be considered as a medium, like a tree bark, communicating through neuron connectivity fed by the input zone. Each tree has lower-resolution branches, which host longer-term memories; higher-resolution branches, which host the short-term memories; and the highest-resolution branch of synapse configurations as working memory. The leaves are the synchronized firings, creating images.

Winds impose stresses on branches, and these stresses have a distribution of strains over branches. The sensory inputs can be given as an example of the tree nutrition from sensory channels in the body as its roots. Winds on the leaves (synapses configurations) and the surface of the tiny branches (high-resolution neural networks) impose strains over some branches (a pathway).

The branches refer back to their original positions when there is no more wind. However, the synapse configurations will not return immediately to their original positions as trees because the neuron fibers and networks are covered in an elastic material, and removing the stresses will not cause them to be returned immediately to the original positions, but the return process is time dependent.

The wind strains the branches and is introduced to an unmet satisfaction when higher humidity or sunshine may make the branches and leaves grow faster and may overmeet the tree's needs. These states of unmet and overmet conditions may not be sensed by a tree, but it initiates a type of emotion for the brain.

The above similarity is useful to provide a frame for basic programming of a part of the dream demonstration package.

Emotions are normally related to the unmet or overmet circumstances of needs and wants.

Needs are mostly physiological, safety, and social; and main desires are the satisfaction of expected esteem and self-actualization.

Emotion in its kinematical definition is a subconscious or conscious brain activity and is usually initiated by a specific sensory input (in waking) or initiated by an overstrained location (in sleep). Emotion typically is accompanied by action, expression, and behavioral changes (in waking) and specific moves—like eyes cycling, male erection, some other behaviors known as disorders, self-expression as dreams, and creations of the new configurations of synapses as new memory of the daily stressful events (during sleep).

Emotion provides a pattern of the synchronized firings, which grow in a self-similar way in energy eddies and energy packets and react with neural networks (sites for energy patterns), creating an energy wake, which provides an image of brain activity like a dream. In waking, the growing self-similar pattern is because of stress flows and their imposing strains; and in sleep, the growing unit pattern of emotion is because of a reverse phenomena of energy relieving from the strained location. Consequently, the energy relief retrieves the related memories by the synchronized firings in the location or by resonating with neural networks. Therefore, an ongoing dream is initiated by an overstrained location as a result of a highly stressful daily event, which is boosted by a resonating long-memory structure to overcome the required activation energy, enabling the location to release the confined energy. If the relief would not happen, the confined energy remains unreleased and trapped on the location, which can cause pain and disorder.

Emotions can be referred to in many ways. However, many psychologists believe that originally, pain and pleasure are the main sources of their development in negative and positive types. Love, joy, and surprise are examples for the pleasure origin; and anger, sadness, and fear are examples for the pain origin. This kind of categorization can be continued in families of emotions, but the main families are not more than about fifty types, like the following:

1.	Affection	20.	Fear	39.	Pride
2.	Anger	21.	Frustration	40.	Rage
3.	Annoyance	22.	Gratitude	41.	Regret
4.	Angst	23.	Grief	42.	Remorse
5.	Apathy	24.	Guilt	43.	Sadness
6.	Anxiety	25.	Happiness	44.	Shame
7.	Awe	26.	Hatred	45.	Shyness
8.	Contempt	27.	Hope	46.	Sorrow
9.	Curiosity	28.	Horror	47.	Suffering
10.	Depression	29.	Hostility	48.	Surprise
11.	Desire	30.	Hysteria	49.	Wonder
12.	Despair	31.	Interest	50.	Worry
13.	Disappointment	32.	Jealousy		
14.	Disgust	33.	Joy	(Reproduced from Wikipedia)	
15.	Ecstasy	34.	Loathing		
16.	Empathy	35.	Love		
17.	Envy	36.	Lust		
18.	Embarrassment	37.	Misery		
19.	Euphoria	38.	Pity		

Emotions can be categorized and linked together in a hierarchical way. Emotions can be categorized and linked together in a hierarchical way.

A detailed alphabetical sorted list can be an example as given in appendix A. The nature of feeling corresponding to each type is not exactly the same for everyone, but it depends on the depth and type of related imposed stress and deviations of a circumstance for the need or desire satisfaction as expected by each individual. The explanation for an emotion for an individual is by character, substance, event, move and behavior, and place and time of related scenario that had created such a circumstance. Different sets of these elements can end in the same feeling of emotion. Therefore, a set of sensory input produce a specific emotion while an emotion can recall or create different scenarios. This makes the difference between types of brain activity in waking and sleep. Therefore, it is said that the characters or behaviors are symbolic in sleep but straight in waking.

Supporting project

To support the kinematics description of the brain activities and the claim that *"dream has a common emotion with 1) a daily stressful event and 2) minimum one previous memory; and 3) it borrows the characters and actions from those events and memories"*, about 185 dreams were noted and analyzed,

1. Analysis was done according to a format made based on the same principle of having shared emotion theme between the stressful daily event, a relevant long memory and the emotion experienced in the dream. Emotion in dream can be the same as well as opposite to the one experienced in the daily event,

2. The emotions were detailed for the dream, the relevant stressful event and the relevant past memories as well as coded according to Hall and Castle Coding System in Ager, Apprehension, Sadness, Confusion and Happiness as general categories.
3. It was found that an stressful event will be followed by a same emotion dream which would happen within a period of a few days.
4. Most of the dreams are symbolic and few are straight reflecting an event, a distorted statement of knowledge or fact,
5. Satisfaction of an unmet need as the sources of high stress and consequently overstrain can be clearly traced in the circumstances that dream narrative provides, with the positive (fulfilling the unmet needs) as well as negative (high lighting the unmet needs) type of emotion.

It was described before that according to Moslow's theory: "people are *motivated to satisfy unmet needs*", including: physical; safety; social; esteem and self-actualization. When a more basic need is satisfied, then a *higher level of need would emerge in brain to be satisfied*. All needs are attractors for the pathway a brain activity finds. **The** *resulting images and thoughts in brain should initiate those motivations*. The motivation in sleep appears as dreaming.

Basic needs are more inner brain oriented, while advance ones are more cortex and front lob oriented. Especially social, esteem and self-actualization needs are associated with self-prestige. Attacking self-prestige needs in waking are highly stressful.

If any of these needs, instead of being satisfied, would be ignored or prevented or attacked, even by oneself, it imposes high stress patterns over the brain. The overall hierarchy of importance is the same as above, but they may make a complex of needs as an attractor to direct the way images to be created.

The criteria for recording and analyzing the dream:

- Date:
- Content of dream:
- Dream analysis:

    - Static/Declarative (substances, peoples, words, . . .):
    - Dynamic/Non-declarative (reasoning, feeling, . . .):
    - Circumstance versus needs (emotion):

- The space which dream happens in:
- Relevant recent events:
- Recent events analysis:

    - Static (substances or peoples):
    - Dynamic (actions, behaviors):
    - Circumstances versus needs (emotion):

- Significant unmet needs (physical; safety; social; esteem and self-actualization) in the dream:
- Relevant past memory:
- Memory analysis:

    - Static (substances or peoples):
    - Dynamic (actions, behaviors):
    - Circumstances (emotion):

- Equivalence table (of static, dynamics, and circumstances) in past memory, diary event and dream:
- Was it an early night dream or midnight dream or early morning dream? (very high stressful dreams like nightmares are expected to happen earlier in sleep periods):
- Does the analysis confirm the formula of Stressful recent events will be resonated with emotion in the relevant past memory creating the emotion felt in the dream scenario? If yes, how strength is the confirmation?

Conclusion of the analysis:

The rating showed that from 177 dreams: 43 dreams confirmed the formula very highly; 76 highly; 41 dreams moderately and 17 dreams cannot be rated or to some extend.

The analysis highly confirms that a dream is a pop up safety valve for high strained tissues which had been strained recently in course of a stressful event and helped to pop up by reinforcing of supporting memories of the same theme emotions.

Appendix A

A Sample of the Detailed Alphabetical Sorted List that can be classified for secondary, tertiary and higher detailed level of emotions

Abandoned	Aggravated	Anguish
Abhor	Aggressive	Anguished
Ablaze	Agitated	Animated
Abominable	*Agonize*	Annoyed
Abrasive	Agonize	Anxiety
Absorbed	Agonized	Anxious
Absurd	Agony	Apathetic
Abused	Agreeable	Apathy
Abusive	Airy	Appealing
Accommodating	Alarmed	Appeasing
Acknowledged	Alienated	Appetizing
Acquiescent	Alive	Appreciation
Acrimonious	Alluring	Apprehensive
Admonished	Alone	Ardent
Adoration	Altruistic	Arduous
Adored	Amazed	Argumentative
Adventurous	Ambiguous	Armored
Adverse	Ambitious	Aroused
Affected	Amenable	Arrogant
Affectionate	Amorous	Astounded
Afflicted	Amused	Attentive
Affronted	Anger	Avid
Afraid	Angry	Avoidance

Awkward
Beaten down
Bemused
Betrayed
Bewildered
Bewilderment
Bewitched
Bitchy
Bitter
Blah
Blessed
Blissful
Blunt
Boiling
Bored
Bothered
Brave
Breathless
Breezy
Bright
Broken
Bruised
Buoyant
Burdened
Burdensome
Bursting
Callous
Calm
Captivated
Captivating
Careless
Caring
Cautious
Celebrating
Chagrined
Charmed
Charming
Chastened

Cheerful
Cherishing
Childish
Clandestine
Clear
Cold
Cold-blooded
Collected
Comatose
Comfortable
Copacetic
Compassion
Competitive
Complacent
Composed
Concern
Concerned
Concerned/worry
Confident
Confused
Congenial
Contemplative
Contempt
Content
Cool
Coping
Cordial
Cornered
Crazy
Creative
Crucified
Crushed
Cursed
Cushy
Cut down
Cynical
Dainty
Defensive

Dejected
Delectable
Delicate
Delighted
Demure
Depressed
Depression
Desirable
Desired
Desolate
Despair
Despondent
Destructive
Devoted
Devoured
Disappointed
Discomfort
Discontent
Discontented
Disgust
Disgust
Disgusted
Dismal
Dispassionate
Displeased
Disappointment
Disregard
Disregarding
Distracted
Distressed
Disturbed
Doldrums
Don't mind
Doomed
Doubtful
Droopy
Dull
Eager

Earnest
Easy
Ecstatic
Electric
Embarrassment
Empathy
Enchanted
Endearing
Enduring
Engaging
Enjoy
Enlivened
Enraged
Enraptured
Enthused
Enthusiastic
Enticing
Envy
Even tempered
Exacerbated
Exasperated
Excited
Excited
Exciting
Exhausted
Exultation
Fanatical
Fascinated
Fascinating
Fear
Fearful
Fearing
Fears of anything
Fervent
Fervor
Fidgety
Fiery
Flared up

Flattering
Flushed
Flustered
Fluttery
Foaming at the mouth
Foolish
Forbearance
Fortitude
Frantic
Fretful
Frigid
Frisky
Frustrated
Frustration
Frustration
Full
Fuming
Fun
Funny
Furious
Galvanized
Gay
Genial
Giddy
Giggly
Glad
Glee
Gleeful
Gloom
Gloomy
Glowing
Gluttonous
Gnawing
Good
Goodness
Grateful
Gratified
Gratitude

Grave
Greed
Greedy
Grief
Grief
Grieving
Grim
Griped
Grounded
Guilt
Guilty
Gushing
Gusto
Haggard
Half-hearted
Happy
Harassed
Hardened
Harsh
Hate
Having Fun
Hearty
Heavy
Hectic
Hesitant
Hilarious
Hope
Hopeful
Horrific
Horrified
Horror-stricken
Hostile
Humorous
Hurt
Hysterical
Idiotic
Ignored
Impatient

Impetuous
Imposing
Impressed
Impressionable
Impulsive
In a dither
In a flurry
In a pickle
In a stupor
In a trance
In purgatory
Inattentive
Indifferent
Indulged
Indulgent
Inept
Infelicitous
Inflexible
Infuriated
Insatiable
Insensitive
Insouciant
Inspired
Interested
Intimidated
Intimidated
Intrigued
Inviting
Irrepressible
Irritated
Irritation
Isolated
Jaunty
Jealous
Jittery
Jolly
Jovial
Joy

Joyful
Jubilation
Jumpy
Keen
Languid
Languish
Laugh
Laughingly
Less than self
Lethargic
Lighthearted
Listless
Lively
Livid
Loathe
Lonely
Lonesome
Long-suffering
Loneliness
Lost
Love
Loved
Loving
Lukewarm
Lust
Luxurious
Mad
Manic
Manipulated
Martyr
Meddlesome
Melancholy
Mellow
Melodramatic
Merry
Mindful
Mindless
Mirthful

Miserable
Moderate
Mopey
Morose
Mortified
Moved
Nervous
Nonchalant
Nostalgic
Not caring
Numb
Obnoxious
Optimistic
Over the edge
Overflowing
Overwhelmed
Overwrought
Pain
Panic
Panic
Paralyzed
Passionate
Passive
Patient
Peace of mind
Perky
Perplexed
Perturbation
Perturbed
Petrified
Phobic
Pain
Pine
Piquant
Pitied
Pity
Placid
Plagued

Pleasant
Pleasing
Pleasurable
Pleasured
Pressured
Prey to
Pride
Protected
Proud
Provocative
Provoked
Quarrelsome
Quenched
Quiet
Quivering
Quivery
Radiant
Rage
Rash
Ravenous
Raving
Ravished
Ravishing
Ready to burst
Receptive
Reckless
Reconciled
Refreshed
Rejected
Rejection
Rejoice
Relish
Remorse
Repressed
Repugnant
Resentful
Resentment
Resigned

Resistant
Restrained
Restraint
Revenge
Revived
Ridiculous
Romantic
Rueful
Sad
Sadness
Safe
Satiated
Satisfaction
Satisfied
Scared
Secretive
Secure
Sedate
Seduced
Seductive
Seething
Selfish
Sensational
Sensual
Sentimental
Serious
Seriousness
Shaken
Shame
Shielded
Shocked
Shutter
Shy
Shyness
Silly
Simmering
Sincere
Sinking

Smug
Snug
Sober
Sobering
Soft
Solemn
Somber
Sore
sorrow
Sorrowful
Sour
Sparkling
Spastic
Spicy
Spirited
Spry
Stoic
Stranded
Stress
stress
Stressed
Stricken
Stung
Stunned
Subdued
Subjugated
Suffering
Sunny
Supportive
Surprise
Surrender
Susceptible
Suspended
Suspicious
Sweet
Sympathy
Taken advantage of
Tame

Tantalizing
Tantrum
Temperate
Tender
Thankful
The blues
Thick-skinned
Thin-skinned
Threatened
Thrill
Thrilled
Tickled
Tight
Tight-lipped
Timid
Tingly
Tired
Tolerant
Tormented
Tortured
Touched
Tranquil
Transported
Trepidation
Troubled
Twitchy
Ugly
Unappeasable
Uncomfortable
Unconcerned
Unconscious
Uncontrollable
Under pressure
Undone
Uneasy
Unfeeling
Unhappy
Unimpressed

Unruffled
Unwise
Used
Vexed
Victim
Victimized
Vivacious
Volcanic
Voluptuous
Vulnerable
Warm
Warmhearted
Wary
Wasteful
Weary
Welcomed
Whining
Wild
Winsome
Wistful
Woe
Woeful
Worked up
Worried
Worrisome
worry
Wounded
Wretched
Yearn
Yearning
Yielding
Zeal
Zealous

Appendix B: An Emotion Tree (as an example)

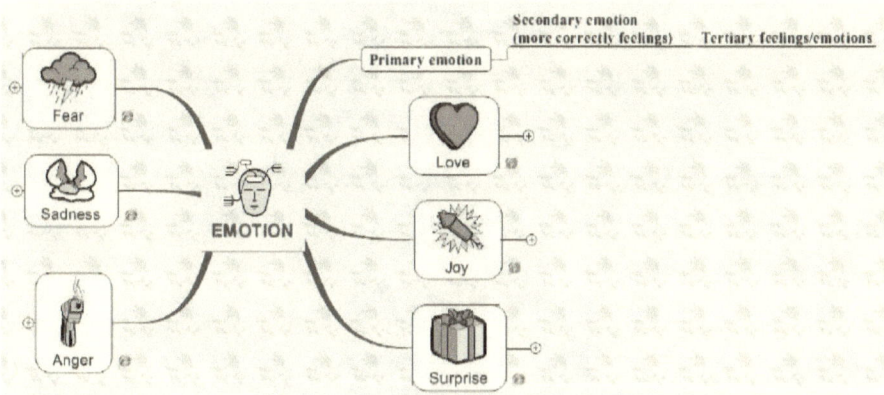

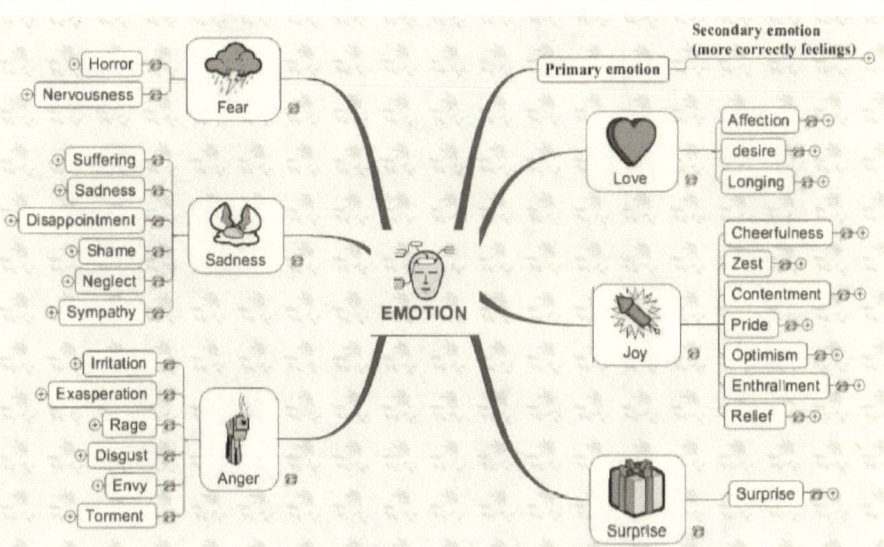

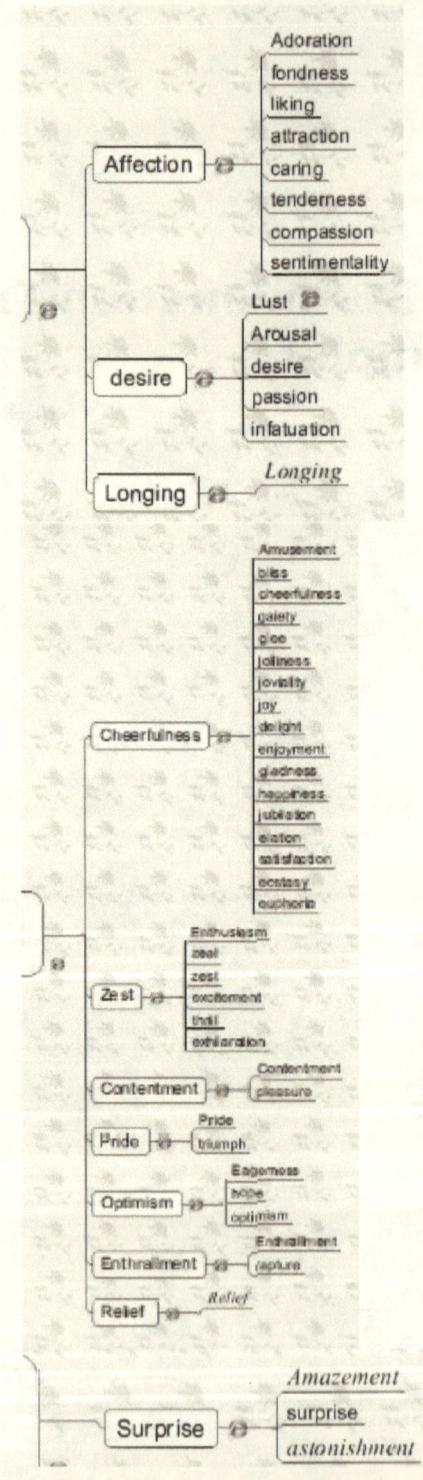

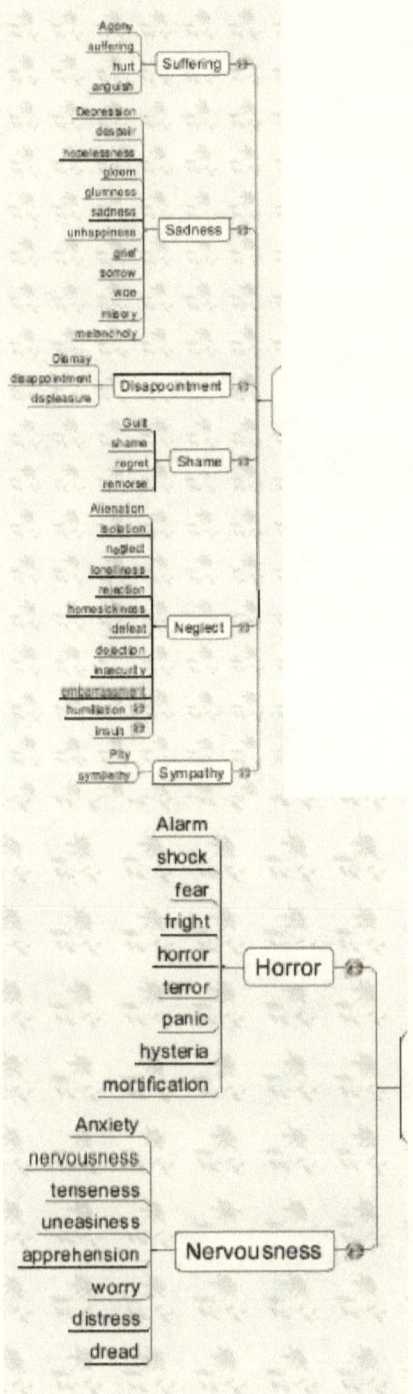

Index

A

abstraction, 15, 78, 109–10
achievers, 31
Active pathways, 69
amygdala, 68, 82, 97, 108, 115
attractors, 13, 15, 18, 22, 46, 77, 109
automade narrative, 80

B

Bilston, Lynne E.
  Neural Tissue Biomechanics, 50
brain
  mapping, 75
  nutrition, 63

C

Cavanna, Andrea E.
  The Precuneus, 39
challenging dreamers, 31
character distinctions, 105
circumstance
  and emotions, 19
  psychological, 53
  real, 19, 53, 113
  virtual, 20, 53, 113
Common Sense, 65–66
"Computing Methods Used to Study 'Servers' in Human Brain" (McIlroy), 38
connectivity, 38, 67
copy-making process, 73

D

delights, 27–28, 100
desires, 13, 31, 115, 117
diary memory element, 47
drags, 36–37, 83
DreamPack, 94
dreams, 9, 20, 34, 70, 81, 92, 99, 101, 103, 112, 114
  logic, 86, 91
  mechanism, 97
  monitoring, 25, 116
  sequence, 31, 33

133

E

eddy, 35–37, 47
Emotion Generator, 94
emotions, 12, 83, 113, 115, 117–18
  categorization of, 53, 102
  and circumstances, 19
  definition of, 23, 97
  development of, 52
  and dreams, 114
  and symbols, 81
energy
  density, 36
  eddies, 35, 117
  fractal, 48
  wake, 23, 48, 73, 117

F

firing fractal, 11, 47, 61, 66, 82, 91–93, 96
free energy
  flow, 18, 34–36, 60, 62–63, 82
  pathway, 48
  transfer, 12–13, 34, 48, 52, 62, 85, 92

G

gap, 41, 43

I

images
  cognition of, 107
  types of, 31
input-output zone, 38, 77–78, 87, 93, 109–10

L

language zone, 78, 109–10
layer tensions, 47

M

Mapping Memory Locations, 70, 75
mathematics, description, 104
McIlroy, Anne
  "Computing Methods Used to Study 'Servers' in Human Brain," 38
memory, 20
  consolidation, 48, 68, 97
  declarative, 20, 33, 53, 68
  fragments, 75
  nondeclarative, 20
mentation, 21, 24, 28, 115
metaphors, 42, 47, 67, 78, 81, 88–89, 99–100, 110
mind, 12, 15, 20, 66, 84, 124, 126

N

Neural Tissue Biomechanics (Bilston), 50
NREM. *See* REM
nutrition emery, 63

O

overstraining, 74, 110

P

pain
  origin, 118
  and pleasure, 16, 53, 118, 126
passive pathway, 69
pleasure. *See* pain and pleasure
Precuneus, The (Cavanna and Trimble), 39
psychological circumstance, 53

R

reality, 84
regularized streams, 37
releasing process, 22, 77, 92, 96
REM, 26, 42–43, 46, 52, 81–82
required activation energy, 40

S

same-emotional memory, 54
selective precuneal hypometabolism, 39
self-expression, 66–67, 100, 117
similes, 47, 67, 78, 81, 89, 99–100, 110
sleep, 13, 20, 23, 34, 42, 46, 85, 89, 98, 113, 119
speed, 40
static memory-saving plans, 108
strain
  patterns, 29, 67
  trend, 56
strain-releasing force, 45
stress
  convection, 36
  diffusivity, 36
stress-strain, 52, 71, 104
superconsciousness, 31
symbols, 52, 66, 78, 81, 97, 109, 114
synapse
  configuration, 54
  level, 23, 33–34, 42, 47, 59, 82, 88–89, 112

T

Trimble, Michael R.
  The Precuneus, 39

W

wakes, 36–37, 73, 83–84

www.ingramcontent.com/pod-product-compliance
Lightning Source LLC
Chambersburg PA
CBHW030808180526
45163CB00003B/1186